PIPEFITTER'S LICENSING
STUDY GUIDE

PIPEFITTER'S LICENSING STUDY GUIDE

Dr. Mark R. Miller
Professor and Chair, Department of Technology
The University of Texas at Tyler

Dr. Rex Miller
Distinguished and Professor Emeritus
Buffalo State College

New York Chicago San Francisco Athens London
Madrid Mexico City Milan New Delhi
Singapore Sydney Toronto

Library of Congress Control Number: 2021931300

Pipefitter's Licensing Study Guide

1 2 3 4 5 6 7 8 9 LWI 26 25 24 23 22 21

ISBN 978-1-260-45826-8
MHID 1-260-45826-1

The pages within this book were printed on acid-free paper.

Sponsoring Editor
Ania Levinson

Editorial Supervisor
Stephen M. Smith

Production Supervisor
Lynn M. Messina

Acquisitions Coordinator
Elizabeth M. Houde

Project Manager
Parag Mittal, KnowledgeWorks Global Ltd.

Copy Editor
KnowledgeWorks Global Ltd.

Proofreader
KnowledgeWorks Global Ltd.

Indexer
Sheila Flavel

Art Director, Cover
Jeff Weeks

Composition
KnowledgeWorks Global Ltd.

This book is dedicated to

Dr. Rex Miller, *as this was the last book he was writing before he passed at the age of 90.*

He authored over 100 books during his lifetime and was an educator for well over 40 years.

Dr. Miller was an inspiration to all those who had the pleasure to meet him and will always be

remembered as a kind, warm-hearted person who would go out of his way to help everyone he met.

About the Authors

Mark R. Miller is a Professor and Chairman of the Technology Department at The University of Texas at Tyler. He has authored over 40 books, including *Electricity and Electronics for HVAC* and *Miller's Guide to Home Plumbing*.

Rex Miller was a Professor Emeritus of Industrial Technology at The State University of New York College at Buffalo. He taught technical courses at all levels for more than 40 years and authored over 100 books, including *Carpentry and Construction*, Sixth Edition, and *HVAC Licensing Study Guide*, Third Edition.

Contents

Preface

Pipefitters are responsible for making, installing, and maintaining piping for primarily high-pressure, commercial applications that involve heating systems, hydraulics and pneumatics, steam boilers, and more. They must know how to cut and weld pipe, read blueprints, and work with valves, flanges, and fittings, as well as understand basic plumbing and rigging in order to move and install large piping systems.

Since states develop their own standards for individuals to become pipefitters, this book only covers the majority of topics that are covered on most pipefitting certification exams across the country. It should be noted that some states also require pipefitters to earn an OSHA 30 card in General Industry safety and a certain number of years as an apprentice learning about the trade before taking a certification exam. Although some safety measures are covered, it is not the intent of this book to cover all of the OSHA standards regarding General Industry training. Furthermore, since some states require individuals to have at least 200 hours of training on ASME standards and codes related to piping systems, this book will touch on it somewhat by providing charts related to these standards and codes.

The intent of this book is to cover the main content areas that are seen on most pipefitting certification exams so individuals can be better prepared for the written portion of these exams. The authors do not imply that this is the only content that is covered in your specific state for pipefitting; however, this book should provide a good review for those taking the certification exam. We wish you the best and hope you have a very rewarding and successful career in pipefitting.

Mark R. Miller
Rex Miller

Acknowledgments

No author works without being influenced and aided by others. Every book reflects this fact. This book is no exception. A number of people cooperated in providing technical data and illustrations. For this we are grateful. We would also like to thank Brandy Nicole Smith Hank, RaeJean Griffin, and others who have contributed to the completion of this book, especially those organizations that so generously supplied information and illustrations.

Those who have been very helpful and cooperative are listed below:

101 Knots
American Welding Society
City of Reno, NV, Public Works Maintenance & Operations
Crosby Group
E.H. Wachs
Fernco
Hoist and Crane Depot
Kawasaki Steel
Lincoln Electric Co.
Linde Div. of Union Carbide
Mazzella Companies
Metabo
NIBCO
Oatey
Red-D-Arc
Stannol
Stevenson Crane Services, Inc.
Texas Vessels & Fabrication, LLC
Turner Div. of Cleanweld Products, Inc.
Victor Equipment Company
Werner Sölken

Techniques for Studying and Test-Taking

PREPARING FOR THE EXAM

1. **Make a study schedule.** Assign yourself a period of time each day to devote to preparation for your exam. A regular time is best, but the important thing is daily study.

2. **Study alone.** You will concentrate better when you work by yourself. Keep a list of questions you find puzzling and points you are unsure of to talk over with a friend who is preparing for the same exam. Plan to exchange ideas at a joint review session just before the test.

3. **Eliminate distractions.** Choose a quiet, well-lit spot as far as possible from telephone, television, and family activities. Try to arrange not to be interrupted.

4. **Begin at the beginning.** Read and underline points that you consider significant. Make marginal notes. Flag the pages that you think are especially important with little Post-it™ notes.

5. **Concentrate on the information and instruction chapters.** Study the code definitions, the *Dictionary of Welding Terms*, and the *Scrambled Dictionary of Equipment and Usage*. Learn the language of the field. Focus on the technique of eliminating wrong answers. This information is important to answering all multiple-choice questions.

6. **Answer the practice questions chapter by chapter.** Take note of your weaknesses; use all available textbooks to brush up.

7. **Try the previous exams, if available.** When you believe that you are well prepared, move on to these exams. If possible, answer an entire exam in one sitting. If you must divide your time, divide it into no more than two sessions per exam. When you do take the practice exams, treat them with respect. Consider each as a dress rehearsal for the real thing. Time yourself accurately and do not peek at the correct answers. Remember, you are taking these for practice; they will not be scored; they do not count. So learn from them.

IMPORTANT: *Do not memorize questions and answers.* Any question that has been released will not be used again. You may run into questions that are very similar, but you will not be tested with these, exact questions. These questions will give you good practice, but they will not have the exact answers to any of the questions on your exam.

HOW TO TAKE AN EXAM

Get to the examination room about 10 minutes ahead of time. You'll get a better start when you are accustomed to the room. If the room is too cold, too warm, or not well ventilated, call these conditions to the attention of the person in charge. Make sure that you read the instructions carefully. In many cases, test-takers lose points because they misread some important part of the directions (e.g., reading the incorrect choice instead of the correct choice).

Don't be afraid to guess. The best policy is, of course, to pace yourself so that you can read and consider each question. Sometimes this does not work. Most civil service exam scores, for instance, are based only on the number of questions answered correctly. This means that a wild guess is better than a blank space. There is no penalty for a wrong answer, and you just might guess right. If you see that time is about to run out, mark all the remaining spaces with the same answer. According to the law of averages, some will be right. Keep in mind that you have bought this book for practice in answering questions.

Part of your preparation is learning to pace yourself so that you need not answer randomly at the end. An educated guess is far better than a wild guess. You make this kind of guess not when you are pressed for time, but when you are not sure of the correct answer. Usually, one or two of the choices are obviously wrong. Eliminate the obviously wrong answers and try to reason among those remaining. Then, if necessary, guess from the smaller field. The odds of choosing a right answer increase if you guess from a field of two instead of from a field of four. When you make an educated guess or a wild guess in the course of the exam, you might want to make a note next to the question number in the test booklet. Then, if there is time, you can go back for a second look. Remember to reason your way through multiple-choice questions very carefully and methodically.

MULTIPLE-CHOICE TEST-TAKING TIPS

Here are a few examples that we can "walk through" together.

I.   On the job, your supervisor gives you a hurried set of directions. As you start your assigned task, you realize you are not quite clear on the directions given to you. The best action to take would be to:

   (a) Continue with your work, hoping to remember the directions.

   (b) Ask a coworker in a similar position what he or she would do.

   (c) Ask your supervisor to repeat or clarify certain directions.

   (d) Go on to another assignment.

In this question you are given four possible answers to the problem described. Though the four choices are all possible actions, it is up to you to choose the best course of action in this particular situation.

Choice (a) will likely lead to a poor result; given that you do not recall or understand the directions, you would not be able to perform the assigned task properly. Keep choice (a) in the back of your mind until you have examined the other alternatives. It could be the best of the four choices given.

Choice (b) is also a possible course of action, but is it the best? Consider that the coworker you consult has not heard the directions. How could he or she know? Perhaps his or her degree of incompetence is greater than yours in this area. Of choices (a) and (b), the better of the two is still choice (a).

Choice (c) is an acceptable course of action. Your supervisor will welcome your questions and will not lose respect for you. At this point, you should hold choice (c) as the best answer and eliminate choice (a). The course of action in choice (d) is decidedly incorrect because the job at hand would not be completed.

Going on to something else does not clear up the problem; it simply postpones your having to make a necessary decision. After careful consideration of all choices given, choice (c) stands out as the best possible course of action.

You should select choice (c) as your answer. Every question is written about a fact or an accepted concept. The question above indicates the concept that, in general, most supervisory personnel appreciate subordinates questioning directions that may not have been fully understood. This type of clarification precludes subsequent errors on the part of subordinates. On the other hand, many subordinates are reluctant to ask questions for fear that their lack of understanding will detract from their supervisor's evaluation of their abilities.

The supervisor, therefore, has the responsibility of issuing orders and directions in such a way that subordinates will not be discouraged from asking questions. This is the concept on which the sample question was based. Of course, if you were familiar with this concept, you would have no trouble answering the question. However, if you were not familiar with it, the method outlined here of eliminating incorrect choices and selecting the correct one should prove successful for you. We have now seen how important it is to identify the concept and the key phrase of the question. Equally, or perhaps even more important, is identifying and analyzing the *key word* or the *qualifying word* in a question—this word is usually an adjective or adverb. Some of the most common key words are:

most	*least*	*best*	*highest*
lowest	*always*	*never*	*sometimes*
most likely	*greatest*	*smallest*	*tallest*
average	*easiest*	*most nearly*	*maximum*
minimum	*only*	*chiefly*	*mainly*
but	*or*		

Identifying these key words is usually half the battle in understanding and, consequently, answering all types of exam questions.

Now we will use the elimination method on some additional questions.

II.   On the first day you report for work after being appointed as an AC mechanic's helper, you are assigned to routine duties that seem to you to be petty in scope. You should:

    (a) Perform your assignment perfunctorily while conserving your energies for more important work in the future.

    (b) Explain to your superior that you are capable of greater responsibility.

    (c) Consider these duties an opportunity to become thoroughly familiar with the workplace.

    (d) Try to get someone to take care of your assignment until you have become thoroughly acquainted with your new associates.

Once again we are confronted with four possible answers from which we have to select the best one.

Choice (a) will not lead to getting your assigned work done in the best possible manner in the shortest possible time. This would be your responsibility as a newly appointed AC mechanic's helper, and the likelihood of getting to do more important work in the future following the approach stated in this choice is remote. However, since this is only choice (a), we must hold it aside because it may turn out to be the best of the four choices given.

Choice (b) is better than choice (a) because your superior may not be familiar with your capabilities at this point. We therefore should drop choice (a) and retain choice (b) because, once again, it may be the best of the four choices. The question clearly states that you are newly appointed. Therefore, would it not be wise to perform whatever duties you are assigned in the best possible manner? In this way, you would not only use the opportunity to become acquainted with procedures, but also to demonstrate your abilities.

Choice (c) contains a course of action that will benefit you and the location in which you are working because it will get needed work done. At this point, we drop choice (b) and retain choice (c) because it is by far the better of the two. The course of action in choice (d) is not likely to get the assignment completed, and it will not enhance your image to your fellow AC mechanic's helpers.

Choice (c), when compared to choice (d), is far better and therefore should be selected as the *best* choice. Now let us take a question that appeared on a police officer's examination.

III.   An off-duty police officer in civilian clothes riding in the rear of a bus notices two teenage boys tampering with the rear emergency door. The most appropriate action for the officer to take is to:

   (a) Tell the boys to discontinue their tampering, pointing out the dangers to life that their actions may create.

   (b) Report the boys' actions to the bus operator and let the bus operator take whatever action is deemed best.

   (c) Signal the bus operator to stop, show the boys the officer's badge, and then order them off the bus.

   (d) Show the boys the officer's badge, order them to stop their actions, and take down their names and addresses.

Before considering the answers to this question, we must accept that it is a well-known fact that a police officer is always on duty to uphold the law even though he or she may be technically off duty.

In choice (a), the course of action taken by the police officer will probably serve to educate the boys and get them to stop their unlawful activity. Since this is only the first choice, we will hold it aside.

In choice (b), we must realize that the authority of the bus operator in this instance is limited. He can ask the boys to stop tampering with the door, but that is all. The police officer can go beyond that point. Therefore, we drop choice (b) and continue to hold choice (a).

Choice (c) as a course of action will not have a lasting effect. What is to stop the boys from boarding the next bus and continuing their unlawful action? We therefore drop choice (c) and continue to hold choice (a).

Choice (d) may have some beneficial effect, but it would not deter the boys from continuing their actions in the future.

When we compare choice (a) with choice (d), we find that choice (a) is the better one overall, and therefore it is the correct answer.

The next question illustrates a type of question that has gained popularity in recent examinations and that requires a two-step evaluation. First, the reader must evaluate the condition in the question as being "desirable" or "undesirable." Once the determination has been made, we are then left with making a selection from two choices instead of the usual four.

IV.   A visitor to an office in a city agency tells one of the office aides that he has an appointment with the supervisor of the office who is expected shortly. The visitor asks for permission to wait in the supervisor's private office, which is unoccupied at the moment. For the office aide to allow the visitor to do so would be:

   (a) Desirable; the visitor would be less likely to disturb the other employees or to be disturbed by them.

   (b) Undesirable; it is not courteous to permit a visitor to be left alone in an office.

   (c) Desirable; the supervisor may wish to speak to the visitor in private.

   (d) Undesirable; the supervisor may have left confidential papers on the desk.

First of all, we must evaluate the course of action on the part of the office aide of permitting the visitor to wait in the supervisor's office as being very undesirable. There is nothing said of the nature of the visit; it may be for a purpose that is not friendly or congenial. There may be papers on the supervisor's desk that he or she does not want the visitor to see or to have knowledge of. Therefore, at this point, we have to decide between choices (b) and (d). This is definitely not a question of courtesy. Although all visitors should be treated with courtesy, permitting the visitor to wait in the supervisor's office in itself is not the only possible act of courtesy. Another comfortable place could be found for the visitor to wait.

Choice (d) contains the exact reason for evaluating this course of action as being undesirable, and when we compare it with choice (b), choice (d) is far better.

A STRATEGY FOR TEST DAY

On the exam day assigned to you, allow the test itself to be the main attraction of the day. Do not squeeze it in between other activities. Arrive rested, relaxed, and on time.

In fact, plan to arrive a little bit early. Leave plenty of time for traffic tie-ups or other complications that might upset you and interfere with your test performance.

Here is a breakdown of what occurs on examination day and tips on starting off on the right foot and preparing to start your exam:

1. In the test room, the examiner will hand out forms for you to fill out and will give you the instructions that you must follow in taking the examination. Note that you must follow instructions exactly.

2. The examiner will tell you how to fill in the blanks on the forms.

3. Exam time limits and timing signals will be explained.

4. Be sure to ask questions if you do not understand any of the examiner's instructions. You need to be sure that you know exactly what to do.

5. Fill in the grids on the forms carefully and accurately. Filling in the wrong blank may lead to loss of veterans' credits to which you may be entitled or to an incorrect address for your test results.

6. Do not begin the exam until you are told to begin.

7. Stop as soon as the examiner tells you to stop.

8. Do not turn pages until you are told to do so.

9. Do not go back to parts you have already completed.

10. Any infraction of the rules is considered cheating. If you cheat, your test paper will not be scored, and you will not be eligible for appointment.

11. Once the signal has been given and you begin the exam, read every word of every question.

12. Be alert for exclusionary words that might affect your answer—words such as "not," "most," and "least."

MARKING YOUR ANSWERS

Read all the choices before you mark your answer. It is statistically true that most errors are made when the last choice is the correct answer. Too many people mark the first answer that seems correct without reading through all the choices to find out which answer is *best*.

Be sure to read the suggestions below now and review them before you take the actual exam. Once you are familiar with the suggestions, you will feel more comfortable with the exam itself and find them all useful when you are marking your answer choices.

1. Mark your answers by completely blackening the answer space of your choice.

2. Mark only ONE answer for each question, even if you think that more than one answer is correct. You must choose only one. If you mark more than one answer, the scoring machine will consider you wrong even if one of your answers is correct.

3. If you change your mind, erase completely. Leave no doubt as to which answer you have chosen.

4. If you do any figuring on the test booklet or on scratch paper, be sure to mark your answer on the answer sheet.

5. Check often to be sure that the question number matches the answer space number and that you have not skipped a space by mistake. If you do skip a space, you must erase all the answers after the skip and answer all the questions again in the right places.

6. Answer every question in order, but do not spend too much time on any one question. If a question seems to be "impossible," do not take it as a personal challenge. Guess and move on. Remember that your task is to answer correctly as many questions as possible. You must apportion your time so as to give yourself a fair chance to read and answer all the questions. If you guess at an answer, mark the question in the test booklet so that you can find it easily if time allows.

7. Guess intelligently if you can. If you do not know the answer to a question, eliminate the answers that you know are wrong and guess from among the remaining choices. If you have no idea whatsoever of the answer to a question, guess anyway. Choose an answer other than the first. The first choice is generally the correct answer less often than the other choices. If your answer is a guess, either an educated guess or a wild one, mark the question in the question booklet so that you can give it a second try if time permits.

8. If you happen to finish before time is up, check to be sure that each question is answered in the right space and that there is only one answer for each question. Return to the difficult questions that you marked in the booklet and try them again. There is no bonus for finishing early so use all your time to perfect your exam paper.

With the combination of techniques for studying and test-taking as well as the self-instructional course and sample examinations in this book, you are given the tools you need to score high on your exam.

PIPEFITTER'S LICENSING STUDY GUIDE

Chapter 1

INTRODUCTION TO PIPEFITTING

Performance Objectives

After studying this chapter you will (be able to):

1.   Recognize the importance of pipes in early history.

2.   Understand how water was made available in large cities.

3.   Discuss how pipes were used in ancient times.

4.   Know what a journeyman pipefitter works with.

5.   Appreciate how pipes advanced civilization.

Pipefitting plays an important role in everyday life in a civilized society. For example, people live comfortably because they can bathe, wash their clothes, and dispose of waste with the aid of plumbing systems. Present-day life utilizes the services of many pipes and the pipefitters who correctly assembled piping components based on detailed drawings. This includes selecting the correct materials and dimensions (both linear and angular) and observing stated specifications and tolerances.

A pipefitter is a tradesperson who installs, assembles, fabricates, maintains, and repairs mechanical piping systems. A pipefitter usually starts or enters the profession as a helper or apprentice. However, the journeyman pipefitter deals with industrial, commercial, marine, and/or heating-cooling systems on a larger scale. Pipefitters work with hydraulics and pneumatics, steam under pressure, hot water heaters, as well as home heating furnaces and home cooling systems, lubrication systems and industrial production equipment that may require maintenance, and, of course, proper installation. Pipefitters may be called upon to:

- Prepare cost estimates for clients.

- Read blueprints and follow state local building codes.

- Determine material and equipment needed for a job.

- Install pipes and fixtures.

- Inspect and test installed pipe systems and pipelines.

- Troubleshoot malfunctioning systems.

- Repair and replace worn parts.

The movement of liquids and gases through pipe is critical to life, especially in the home, when water is needed for both drinking and sanitation. In factories, chemicals are moved to aid in product manufacturing. In power plants steam is moved to drive turbines that generate electricity. Pipefitters install and repair these pipe systems. See Figures 1.1 and 1.2.

HISTORY OF PIPES

The history of pipes in the United Kingdom stretches back to the Romans, and it serves to enlighten us as to why pipes and piping are necessary for life in a civilized society. Evidence of various degrees of engineering marvels has been found from as far back as 43AD when the Romans invaded England. The Romans were masters of engineering. Through a series of aqueducts, canals, and rolled-lead piping, water was fed from rivers, springs, and lakes to a vast network of towns, cities, and garrisons. This water usually ended up in a central location and was piped to public fountains and baths. It was found in their society that people of highest status who owned large houses had their own connected directly to the water supply.

FIGURE 1.1 Piping systems for a refinery.

FIGURE 1.2 Pipefitter preparing a weld with a grinder.

All through the Roman rule, developments in sanitary care increased and a cleaner way of life was created. During the fall of the Roman Empire in the 6th century AD with the rise of the Barbarians, Picts, Saxons, and Irish, the Romans fled from its northern territories and with them England lost its sanitary sewers and civilization as we know it today. It was clear this regression was inevitable and sanitation became very basic or crude.

Time marched on and Christianity began its rule of England. The new rules were cynical of anything Roman, and they saw bathing as vain and associated Roman bathing houses with debauchery and frowned on their use and upkeep. This view directly impacted the living conditions.

The Black Plague followed with the loss of one third of the population as the rats and fleas enjoyed themselves in the rubbish heaps. It took about five centuries for the living conditions of the population to take a turn for the best with the use of more modern sanitation devices and practices. Previously, safe drinking water was not a reality, so many turned to wine and beer as the drink of choice. Typically, only the wealthy bathed with any regularity at that time. Running water was not common, and only some of the religious houses made progress by building open channels to towns and monasteries. This, again, was rare and done with no real piping. The only progress, if you can call it that, was in water and sanitation throughout the Middle Ages and beyond came in the development of sewage and drainage ditches. These really were little more than gutters in the street or covered tunnels in places built without thought and were aimed toward a river.

It was not until the year 1460 when the city of Hull, England, laid lead pipes throughout the town, and then householders could pay to install a water pump within the home. In 1584, the town of Plymouth, England, installed a water system bringing water to the town and storing it in cisterns for public use free of charge. Oxford built covered gullies to bring spring water to a 20,000-gallon tank for public use.

The Victorian era and the Industrial Revolution brought the biggest change in water movement and pipes with areas of population gaining central pumps, but these pumps were only available for a short time each day. The water was carried to the home in any kind of vessel available. The grander houses were fitted with pipes, but only to one floor, and the water was heated in kettles and moved around the house by hand. It was not until the 1800s that towns and cities such as Manchester, London, Liverpool, and Aberdeen began the building of reservoirs. Leicester built the first sewage treatment works. Bristol gained a pipeline and aqueduct to bring clean water from over 20 miles away. It was during this time the government began to legislate and get interested in drinking water so they made polluting drinking water a criminal offence in 1847. In 1848, the Public Health Act was passed and created the model for the current plumbing code we use today. The act made it mandatory for some form of sanitary unit in every house; this could be a flushing toilet or privy. The government of that day also invested the significant sum of 5 million pounds for research and engineering works, and from here on a solid sewer system grew.

PIPE TECHNOLOGY

The technology for pipes developed slowly from the lead of the Roman period. Iron and wooden piping was then used in the Middle Ages; however, lead was still used for smaller gauge. Wooden piping has been found in homes as recent as the 1890s. The advent of inexpensive cast iron pipe in the 1960s coupled with economical light gauge copper tubing enabled pipefitters and plumbers the ability to fit piping to new and old properties. In 1926, polyvinyl chloride (PVC) pipe was invented, but not manufactured in volume until the 1940s, and was used extensively in the rebuilding of Japan and Germany after World War II. Although copper tubing and pipe are still prevalent for water lines in new construction, the rising price of copper has enabled plastic (PVC and PEX) to become a welcome alternative; refer to Figure 1.3.

FIGURE 1.3 White PVC drain, waste, and vent (DWV) pipe and PEX hot and cold water lines.

REVIEW QUESTIONS

1. In most cases, how does a pipefitter enter the profession?

   a. apprentice
   b. journeyman
   c. master
   d. none of these

2. Which of the following is NOT something a pipefitter would be required to do?

   a. prepare cost estimates
   b. read blueprints and follow state/local building codes
   c. troubleshoot malfunctioning water systems
   d. none of these

3. How are pipes used in power plants to generate electricity?

   a. transport water to turn paddle wheels
   b. transport oil to drive generators
   c. transport steam to drive turbines
   d. none of these

4. Which civilized society is noted for engineering and bringing piping to the United Kingdom?

   a. Visigoths
   b. Greeks
   c. Huns
   d. Romans

5. What was probably a major cause of the Black Plague?

   a. Roman invasion of the United Kingdom
   b. Abandoning Roman culture of bathing, and piped water
   c. Barbarian, Saxon, and Picts use of cast iron pipe
   d. none of these

6. Early piping was made of _____.

   a. lead, wood, and iron
   b. cast iron and clay
   c. copper and lead
   d. none of these

7. Which era brought the biggest change in water movement through pipe?

   a. Renaissance
   b. Victorian
   c. Progressive
   d. Dark Ages

8. When were reservoirs first built in England for residential use?

    a. 1400s   b. 1600s

    c. 1800s   d. 1900s

9. What type of pipe was used extensively to rebuild Japan and Germany?

    a. lead   b. iron

    c. wooden   d. PVC

10. What type of tubing is now becoming prevalent in new construction?

    a. iron   b. brass

    c. copper   d. plastic

ANSWERS TO REVIEW QUESTIONS

1. a	2. d	3. c	4. d	5. b
6. a	7. b	8. c	9. d	10. d

Chapter 2 ———————

METAL PIPE STANDARDS AND IDENTIFICATION

Performance Objectives

After studying this chapter you will (be able to):

1.   List and describe the seven classifications of low-pressure piping systems.

2.   Discuss the organizations created to ensure quality standards for products.

3.   Explain the grading system used for high-pressure pipe.

4.   Understand the differences between pipe and tubing.

5.   Interpret pipe weight and schedule charts to select the correct pipe for a job.

6.   Know how pipe is identified and finished.

CLASSIFICATION OF PIPE

Metal Pipe

Metal pipes are perhaps the oldest type used. Lead pipes were used in Roman times. Cast iron pipes have been in use for over a century. Time and experience have provided some insights into the best uses of metal pipes to avoid such outcomes as lead poisoning. Metal pipes include pipes made from steel, cast iron, copper, lead, and brass.

The pipefitter must be acquainted with a number of types of pipe and know how to identify them when called upon. Pipe has several classifications depending upon if it is used for low- or high-pressure systems. Although plumbers typically work with low-pressure systems, pipefitters need to be aware of how they work in case these systems must be removed and reassembled and/or become damaged when installing a high-pressure system.

Low-Pressure Pipe Classifications

There are seven classifications of pipe materials contained in most city codes for low-pressure systems:

1.   Water-service: This type of pipe may be made of asbestos cement (discontinued in 1992 due to health concerns), brass, cast iron, copper, plastic, galvanized iron, or steel.

2.   Water-distribution pipe: This type of pipe may be made of brass, copper, plastic, or galvanized steel.

3.   Above-ground drainage and vent pipe: This type of pipe may be made of brass, cast iron, type K, L, M, or drain, waste, and vent (DWV) copper, galvanized steel, or plastic.

4.   Underground drainage and vent pipe: This type of pipe can be made of brass, cast iron, type K, L, M, or DWV copper, galvanized steel, or plastic.

5.   Building-sewer pipe: This type of pipe may be made of asbestos cement, bituminized fiber, cast iron, type K or L copper, concrete, plastic, or vitrified clay.

6.   Building-storm-sewer pipe: This type of pipe may be made of asbestos cement, bituminized fiber, cast iron, concrete, type K, L, M, or DWV copper, or vitrified tile.

7.   Subsoil-drain pipe: This type of pipe may be made of asbestos cement, bituminized fiber, cast iron, plastic, styrene rubber, or vitrified clay.

Types of Copper Pipe and Tubing

Copper pipe is manufactured to different specifications than tubing, and its size may not refer to its actual inside diameter (ID) or outside (OD). The wall thickness of the copper pipe changes accordingly with pipe diameter and can be purchased in regular or extra strong grades. For example, a 1-inch copper pipe whether it is standard or extra strong would have the same OD of approximately 1-5/16 inches; however, the wall thickness for the extra strong would be 0.056 inch thicker than that of the regular pipe. This would then mean that the standard pipe could handle more fluid with the thinner wall thickness, but the extra strong could handle higher pressures. Piping is used primarily for commercial jobs that typically require much higher pressures than tubing.

There are several types of copper tubing. One is structural tubing that can come in round, square, rectangular, or any other formed shape; however, it is not meant to carry fluids and its size refers to its actual outside diameter. Another type of tubing is copper water tube and is used for plumbing. It can come in hard or soft form. The straight length that is not really meant for bending is hard tube and has a temper. It is often referred to by plumbers as pipe because it comes in straight lengths as does most other pipe. The tube that has been annealed and sold in rolled up coils is the soft form and is referred to as tubing. Copper water tubing size refers to its inside diameter since it is more concerned with the amount of fluid it can carry.

There are two main types of copper tube. One is used for plumbing and the other is used for air conditioning and refrigeration (ACR) and/or for medical (MED) lines which may also use the abbreviation of OXY for lines that carry oxygen gas. The other reason for this distinction is that the plumbing tubing dimension refers to the inside diameter whereas the ACR/MED refers to its outside diameter. The ACR/MED is also more expensive because it has to be clean of any manufacturing oils and both ends must also be capped to keep it clean before it is used. This is because oxygen is typically piped through ACR/MED copper lines at medical facilities and pressurized oxygen can explode when it comes in contact with grease or oil. Plumbing tube includes types K, L, M, and DWV. Types K, L, M, and DWV come in standard sizes with the outside diameter always ⅛th inch (0.32 cm) larger than the standard size. Each type represents a series of sizes with different wall thicknesses. Inside diameters depend on the tube size and wall thickness. Drawn tube is hard tube and annealed tube is soft. Hard tubing can be joined by soldering or brazing, by using capillary fittings, or by welding. Tubing that may be bent would be annealed and can be joined in the same ways, as well as with flare-type compression fittings. It is also possible to expand the end of one tube so that it can be joined to another by soldering or brazing without a capillary fitting.

Strength, formability, and other mechanical factors frequently determine the type of copper tube to be used in a particular application. Sometimes building or plumbing codes govern what types may be used. When a choice can be made, it is helpful to know which type of copper tube has served and will serve successfully and economically in the following operations:

Type K copper is typically available in straight hard pipe and soft annealed tubing for serious flow of water. It has a thicker wall section than L, M, or DWV. It is typically used for underground water line and commercial applications. The letter marking the pipe is typically green.

Type L copper comes in hard straight pipe and soft annealed tubing as well, however, the wall section is thinner than type K copper. It is typically used in smaller commercial and most residential interior water line applications. The letter marking the pipe is usually printed in blue.

Type M copper is typically available in straight lengths and is thinner than type L copper pipe. It is mainly used for heating applications and not allowed for water use in most cities. This pipe is usually identified with red lettering.

Type DWV copper is the thinnest wall section of copper pipe and only comes in straight pipe. Its primary use is for drains, waste, and venting; however, now that PVC is fire rated, it is not used as frequently because of its comparative cost. The lettering on the pipe is yellow. It is only rated at 15 psi.

High-Pressure Pipe Classifications

In most cases, a pipefitter works with high-pressure systems that require a different set of skills. High-pressure systems are primarily made of steel and require the pipefitter to read piping diagrams, blueprints, weld metal, calculate offsets, and understand and use rigging to move heavy pipe.

In order to ensure the quality of piping, for example, to make sure it will last as long as another manufacturer's pipe that is made to the same diameter and thickness, certain organizations were established to test the products. The American Society for Mechanical Engineers (ASME) has developed various types of standards for pipe and works with the American National Standards Institute (ANSI). Typically, one organization will endorse the others. In addition, the American Society for Testing and Materials (ASTM) will also test and have standards for pipe as does the American Petroleum Institute (API). Therefore, when purchasing pipe for the correct application, it could have several standards listed. For example, a standard steel pipe used for a pipeline could have the following standards listed by it so the purchaser would know that the pipe would meet the requirements for pressure, chemical resistance, etc. that was needed to ensure safe and long-lasting operation; ASTM A252, API 5L, ANSI B31.4, and ASME B31. Depending on the organization, pipe grades will be listed as A, B, C or 1, 2, 3 with A and 1 having the least yield and tensile strength while C and 3 would have higher tensile strengths. In addition, API may designate an X with a number after the standard in lieu of A, B, or C. Therefore, API 5L X70 would designate that the minimum yield strength is the number times 1000 or 70,000 psi instead of the 30,000 psi of grade A. The minimum yield pressure is increased by adding more carbon and manganese to the steel. However, the harder the pipe is, the more brittle it becomes and may not necessarily be a good characteristic in certain applications. In addition, if a liquid or gas is flowing through the pipe at a much lower pressure than a grade A or 1 is rated for, then there is no sense in paying more for pipe that will never require a higher rating. Furthermore, API may use the designation PSL 1 and PSL 2. The PSL stands for product standard level and PSL 2 requires more testing and certification for pipe quality than PSL 1. Refer to Table 2.1 for more information.

Pipe versus Tubing Size

Tubing is sized differently than pipe because its walls are thinner so it can be bent. It was originally constructed to be used to build parts with. Moreover, copper tubing did not become popular for use in plumbing until the 1960s after standards were already put in place for pipe. Pipe on the other hand is concerned with the inside diameter because its purpose is to move liquids and gases. Therefore, tubing sizes refer to outside diameters (OD) when pipe sizes refer to inside diameters (ID); refer to Figure 2.1. It should also be noted that pipe sizes are really sized by their nominal pipe size (NPS) or dimension nominal (DN), which is the international designation. The NPS is in between the outside and inside diameter and changes in respective to the thickness preference for wall size. In other words, pipe can be made in different thicknesses depending upon the line pressures of the pipe, so the higher the pressure, the thicker the wall of the pipe. In 1927, when the pipe standards were originally derived, only a few wall thicknesses were in use: standard weight (STD), extra-strong (XS), and double extra-strong (XXS). In 1939, it was hoped that the designations of STD, XS, and XXS would be phased out by schedule numbers; however, those original terms are still in common use today and now referred to as extra-heavy (XH), and double extra-heavy (XXH), respectively.

PIPE SCHEDULE NUMBERS AND WEIGHTS

As noted in the previous section, the three pipe weights were the original way to designate the various wall thicknesses of pipe or strength of pipe (refer to Figure 2.2). Schedule numbers came into play due to the numerous applications that were evolving for pipe after WWII along with the new ways to manufacture it.

TABLE 2.1 Listing of pipe grades and their ratings.

Grade A Carbon Steel	(ASTM A53, A523, API 5L PSL1)	
Minimum Yield Strength:	30,000 psi	
Maximum Yield Strength:	None	
Minimum Tensile Strength:	48,000 psi	

Notes: This grade may be used for standard pipe, line pipe, or conduit pipe as specified in the various ASTM and API specifications.

Grade B Carbon Steel	(ASTM A53, A523, API 5L PSL1)	API 5L PSL2
Minimum Yield Strength:	35,000 psi	35,000 psi
Maximum Yield Strength:	None	65,000 psi
Minimum Tensile Strength:	60,000 psi	60,000 psi
Maximum Tensile Strength:	None	110,000 psi

Notes: This grade may be used for standard pipe, line pipe, or conduit pipe as specified in the various ASTM and API specifications.

API 5L X42	PSL1	PSL2
Minimum Yield Strength:	42,000 psi	42,000 psi
Maximum Yield Strength:	None	72,000 psi
Minimum Tensile Strength:	60,000 psi	60,000 psi
Maximum Tensile Strength	None	110,000 psi

Notes: This is a standard line pipe grade specified in API (American Petroleum Institute) specification 5L.

API 5L X46	PSL1	PSL 2
Minimum Yield Strength:	46,000 psi	46,000 psi
Maximum Yield Strength:	None	76,000 psi
Minimum Tensile Strength:	63,000 psi	63,000 psi
Maximum Tensile Strength:	None	110,000 psi

Notes: This is a standard line pipe grade specified in API (American Petroleum Institute) specification 5L.

API 5L X52	PSL1	PSL2
Minimum Yield Strength:	52,000 psi	52,000 psi
Maximum Yield Strength:	None	77,000 psi
Minimum Tensile Strength:	66,000 psi	66,000 psi
Maximum Tensile Strength:	None	110,000 psi

Notes: This is a standard line pipe grade specified in API (American Petroleum Institute) specification 5L.

API 5L X56	PSL1	PSL2
Minimum Yield Strength:	56,000 psi	56,000 psi
Maximum Yield Strength:	None	79,000 psi
Minimum Tensile Strength:	71,000 psi	71,000 psi
Maximum Tensile Strength:	None	110,000 psi

Notes: This is a standard line pipe grade specified in API (American Petroleum Institute) specification 5L.

API 5L X56	PSL1	PSL2
Minimum Yield Strength:	56,000 psi	56,000 psi
Maximum Yield Strength:	None	79,000 psi
Minimum Tensile Strength:	71,000 psi	71,000 psi
Maximum Tensile Strength:	None	110,000 psi

Notes: This is a standard line pipe grade specified in API (American Petroleum Institute) specification 5L.

(*Continued*)

TABLE 2.1   Listing of pipe grades and their ratings. (*Continued*)

API 5L X56	PSL1	PSL2
Minimum Yield Strength:	56,000 psi	56,000 psi
Maximum Yield Strength:	None	79,000 psi
Minimum Tensile Strength:	71,000 psi	71,000 psi
Maximum Tensile Strength:	None	110,000 psi

Notes: This is a standard line pipe grade specified in API (American Petroleum Institute) specification 5L.

API 5L X60	PSL1	PSL2
Minimum Yield Strength:	60,000 psi	60,000 psi
Maximum Yield Strength:	None	82,000 psi
Minimum Tensile Strength:	75,000 psi	75,000 psi
Maximum Tensile Strength:	None	110,000 psi

Notes: This is a standard line pipe grade specified in API (American Petroleum Institute) specification 5L.

API 5L X65	PSL1	PSL2
Minimum Yield Strength:	65,000 psi	65,000 psi
Maximum Yield Strength:	None	87,000 psi
Minimum Tensile Strength:	77,000 psi	77,000 psi
Maximum Tensile Strength:	None	110,000 psi

Notes: This is a standard line pipe grade specified in API (American Petroleum Institute) specification 5L.

API 5L X70	PSL1	PSL2
Minimum Yield Strength:	70,000 psi	70,000 psi
Maximum Yield Strength:	None	90,000 psi
Minimum Tensile Strength:	82,000 psi	82,000 psi
Maximum Tensile Strength:	None	110,000 psi

Notes: This is a standard line pipe grade specified in API (American Petroleum Institute) specification 5L.

Grade 1 Carbon Steel	(ASTM A252)
Minimum Yield Strength:	30,000 psi
Maximum Yield Strength:	None
Minimum Tensile Strength:	50,000 psi

Notes: This grade is a standard grade for ASTM A252. The specification covers the requirements for welded and seamless steel piling pipe.

Grade 2 Carbon Steel	(ASTM A252)
Minimum Yield Strength:	35,000 psi
Maximum Yield Strength:	None
Minimum Tensile Strength:	60,000 psi

Notes: This grade is a standard grade for ASTM A252. The specification covers the requirements for welded and seamless steel piling pipe.

Grade 2 Carbon Steel	(ASTM A252)
Minimum Yield Strength:	35,000 psi
Maximum Yield Strength:	None
Minimum Tensile Strength:	60,000 psi

Notes: This grade is a standard grade for ASTM A252. The specification covers the requirements for welded and seamless steel piling pipe.

Grade 3 Carbon Steel	(ASTM A252)
Minimum Yield Strength:	45,000 psi
Maximum Yield Strength:	None
Minimum Tensile Strength:	66,000 psi

Notes: This grade is a standard grade for ASTM A252. The specification covers the requirements for welded and seamless steel piling pipe.

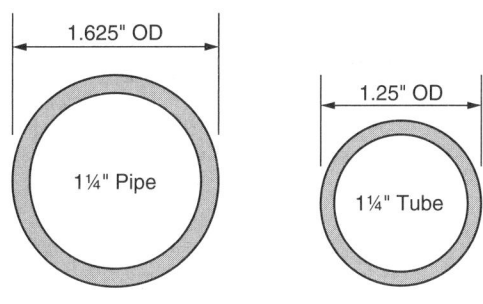

FIGURE 2.1 Outside diameter pipe size comparison of pipe verses tubing.

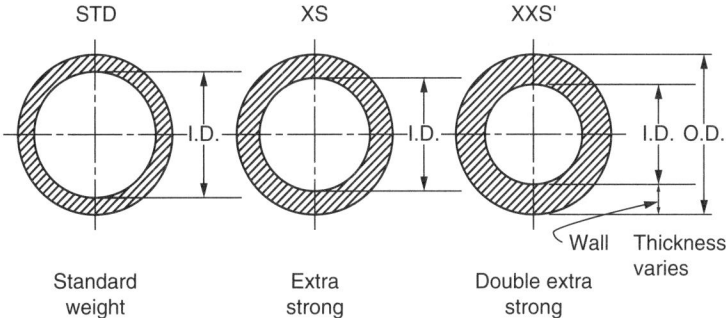

FIGURE 2.2 Illustration of pipe weight wall thicknesses.

However, weights and schedule numbers are still used and they start to overlap, depending upon the diameter of the pipe.

Pipe schedule numbers range from 10 to 160 with the wall thickness of the pipe increasing with the number size. The schedule pipe sizes are 10, 20, 30, 40, 60, 80, 100, 120, 140, and 160. It should be noted that wall thickness for STD pipe and Schedule 40 pipe are the same from ⅛ to 10 inches in diameter. After 10 inches, the STD pipe maintains a wall thickness of ⅜ inch while the scheduled pipe increases proportionally. The XH weight designation is the same as Schedule 80 until over the pipe diameter of 8 inches. After 8 inches the XH designation maintains a ⅛-inch wall thickness while the Schedule 80 increases proportionally to diameter. As for the XXH pipe weight designation, there is no schedule pipe thicker than it until Schedule 160 pipe becomes larger than 6 inches in diameter (see Figures 2.2 and 2.3). When purchasing stainless steel pipe the wall thickness does not have to be as thick so the schedule sizes may differ and start at Schedule 5. Stainless pipe is designated with the letter s after the schedule number as illustrated in Table 2.2.

FIGURE 2.3 Example of same outside diameter pipe in various schedule sizes.

TABLE 2.2 Wall thickness chart for various schedule and weights of varying diameter pipe.

Pipe Schedules Wall Thickness (inches)

Nominal	O.D. inches	5s	5	10s	10	20	30	40s & Std	40	60	80s & E.H.	80	100	120	140	160	Dbl. E.H. (XXH)
⅛	.405		.035	.049	.049			.068	.068		.095	.095					
¼	.540		.049	.065	.065			.088	.088		.119	.119					
⅜	.675		.049	.065	.065			.091	.091		.126	.126					
½	.840	.065	.065	.083	.083			.109	.109		.147	.147				.187	.294
¾	1.050	.065	.065	.083	.083			.113	.113		.154	.154				.218	.308
1	1.315	.065	.065	.109	.109			.133	.133		.179	.179				.250	.358
1¼	1.660	.065	.065	.109	.109			.140	.140		.191	.191				.250	.382
1½	1.900	.065	.065	.109	.109			.145	.145		.200	.200				.281	.400
2	2.375	.065	.065	.109	.109			.154	.154		.218	.218				.343	.436
2½	2.875	.083	.083	.120	.120			.203	.203		.276	.276				.375	.552
3	3.500	.083	.083	.120	.120			.216	.216		.300	.300				.437	.600
3½	4.000	.083	.083	.120	.120			.226	.226		.318	.318					.636
4	4.500	.083	.083	.120	.120			.237	.237	.281	.337	.337		.437		.531	.674
4½	5.000							.247			.355						.710
5	5.563	.109	.109	.134	.134			.258	.258		.375	.375		.500		.625	.750
6	6.625	.109	.109	.134	.134			.280	.280		.432	.432		.562		.718	.864
7	7.625							.301			.500						.875
8	8.625	.109	.109	.148	.148	.250	.277	.322	.322	.406	.500	.500	.593	.718	.812	.906	.875
9	9.625							.342			.500	.500					
10	10.750	.134	.134	.165	.165	.250	.307	.365	.365	.500	.500	.593	.718	.843	1.000	1.125	

PIPE LENGTHS

The standard lengths most manufacturers make steel pipe are 20' and 40'. Steel pipe is also supplied in other lengths that are broken down into one of the four categories:

1. *Single random length* which is from 18 to 25 feet
2. *Double random length* which is from 38 to 40 feet
3. *Longer than double random length* longer than 38 feet and can go up to 80 feet
4. *Cut lengths* which can be cut to any length within ⅛ inch up to 80 feet

STEEL PIPE COATINGS

Steel pipes can be ordered in a multitude of ways in regards to coatings. In most cases, it is coated with a thin, black lacquer finish to prevent it from rusting in transit. However, it can also be ordered with a bare finish, pickled only, pickled and oiled, galvanized (pipe is dipped in hot zinc to prevent it from rusting), or coated with aluminum, chromium, plated with a variety of metals, oiled, painted, or even coated with plastic, cement, rubber, glass, brick, concrete, and even coal tar.

PIPE END FINISH

When purchasing steel pipe, the way the end is finished can be specified. The four main categories are as follows:

1. *Plain ends (PEs)* are when the ends of pipes are cut square as seen in Figure 2.4.
2. *Threaded ends (TEs)* are typically used for pipes with a nominal size of 3-inches or smaller. TE pipes allow for an excellent seal and use the national pipe thread (NPT) standard with the most common taper measuring ¾-inch per foot. Refer to Figure 2.5.
3. *Beveled ends (BWs)* are when the end or ends of the pipe come beveled to a certain angle as shown in Figure 2.6.

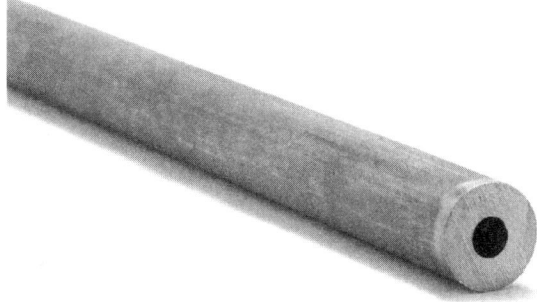

FIGURE 2.4 Pipe with cut plain end.

FIGURE 2.5 Threaded-end finish for pipe.

FIGURE 2.6   Pipes manufactured with beveled ends.

FIGURE 2.7   Steel pipe with grooved ends and half the housing that is used to seal pipes.

4.   *Grooved mechanical joints or grooved ends* use a formed or machined groove at the end of the pipe to seat a gasket. A housing around the gasket is then tightened to secure the connection and ensure optimal seal and performance. The design allows for easier disassembly with a reduced risk of damaging piping components as illustrated in Figure 2.7.

Steel pipe can come with a variety of end finishes as noted by the following abbreviation list:
BE: Bevel End

BBE: Bevel Both Ends

BLE: Bevel Large End

BOE: Bevel One End

BSE: Bevel Small End

BW: Butt-weld End

PBE: Plain Both Ends

PE: Plain End

POE: Plain One End

TBE: Thread Both Ends

TE: Thread End

TLE: Thread Large End

TOE: Thread One End

TSE: Thread Small End

A single pipe can be ordered with two different end types. This is often designated in the manufacturer's pipe description or label. For example, a ¾-inch SMLS Schedule 80s A/SA312-TP316L TOE pipe label would mean that the pipe has threads on one end (TOE) and is plain on the other. In contrast, a ¾-inch SMLS Schedule 80s A/SA312-TP316L TBE pipe would have threads on both ends (TBE).

FIGURE 2.8   An example of manufacturing markings on steel pipe.

PIPE IDENTIFICATION

As noted previously, pipes come in all different grades of steel, diameters, lengths, end finishes, etc., so a pipefitter must be able to read the label correctly in order to select the correct type of pipe for the job. The various pipe manufacturers follow the specifications set forth by the organizations that ensure quality. Therefore, most pipe manufacturers will use the ASTM, API, or the Canadian Standard Association (CSA) specifications criteria on their label. An example of how a manufacturer labels steel pipe is given in Figure 2.8. Most manufacturers provide the following information on steel pipe:

Company name, name of organization's standard that is followed, for example ASTM A53/A106 or API 5L; grade A, B, or C; outside diameter in inches or millimeters; how it was manufactured, for example S = seamless, E = welded, F = butt-welded only, SW = spiral welded; type of steel, E = electric furnace, R = rephosphorized, no marking for open hearth or basic oxygen made steel; heat treatment, HN = normalized, HS = subcritical stress relieved, HA = subcritical age hardened, HQ = quenched and tempered; test pressure, is it higher than in the tables or NH = when not tested, schedule number, S = supplementary requirements; and pipe length.

GALVANIZED PIPE

As shown in Figure 2.9, galvanized pipe is a steel pipe that has been coated with zinc inside and outside to reduce corrosion. Note the word *reduce* in the previous sentence: the coating reduces corrosion; it doesn't stop it. Galvanized pipe may last a very long time, but if you have galvanized supply lines and notice rust stains or a rusty taste to the water, it is because the corrosion is advanced.

BLACK PIPE

Black pipe does not have a coating. It is called black pipe because a black scale is formed on the surface when it is shaped at the steel mill. Sometimes the steel mill will apply a thin coat of black varnish or lacquer to the pipe, as seen in Figure 2.10. This type of pipe is normally used only for gas lines in homes. It will rust out too quickly to be used for water or chemicals. Galvanized pipe may also be used for gas lines but is a bit more expensive.

Elbow
Fig No.90
1/2"-4"

Reducing Elbow
Fig No.90R
1/2"-4"

M&F Elbow
Fig No.92
1/2"-4"

Bushing
Fig No.241
1/2"-4"

Tee
Fig No.130
1/2"-4"

Reducing Tee
Fig No.130R
1/2"-4"

Socket
Fig No.220
1/2"-4"

M&F Socket
Fig No.529
1/2"-4"

Reducing Socket
Fig No.240
1/2"-4"

Beaded Plugs
Fig No.290
1/2"-4"

Plain Plugs
Fig No.291
1/2"-4"

Hexagon Nipple
Fig No.280
1/2"-4"

Cross
Fig No.180
1/2"-4"

Union
Fig No.340
1/2"-4"

Union
Fig No.330
1/2"-4"

Cap
Fig No.301
1/2"-4"

FIGURE 2.9   Galvanized pipe and fittings.

FIGURE 2.10   Black pipe is steel pipe with no corrosion-resistant coating.

CAST IRON PIPE

Cast iron pipes are made of real cast iron. They are tough, durable, strong, and heavy. They are used almost exclusively for main DWV pipes and are not available in smaller sizes appropriate for supply lines. They are difficult to cut and complicated to seal. Many codes now require repairs and replacements to be made with plastic pipe. Table 2.3 shows information about metal pipes.

TABLE 2.3   Metal pipe specifications.

Type	Use	Diameters, inches	Std. Lengths, feet	Joining
Rigid copper	Hot and cold water supply	⅜, ½, ¾, 1	10, 20	Soldered fittings
Flexible copper	Hot and cold water supply and gas lines	¼, ⅜, ½, ¾, 1	30 or 60 coil, 10 and 25 coils in some areas	Flare or compression fittings
Galvanized steel	Hot and cold water supply and DWV	⅛, ¼, ⅜, ½, ¾, 1, 1½, 2	21 Joint	Threaded fittings
Brass	Valves, special drains	¼ to 1½	Varies—special items	Special fittings
Cast iron	Main drains, vents	3, 4	5, 10	Packed and leaded fittings

BRASS PIPE

Brass pipe is commonly used for traps, drains, and connecting lines. It is one of the most expensive types of pipe. For household uses, it is usually chrome-plated for appearance. Brass pipe is heavy, difficult to work with, and expensive. As copper is the main ingredient of brass, copper and brass fittings can be used interchangeably. Brass pipe can be threaded or soldered. The pipes can be sweated in with a torch just like copper pipes are with flux and solder. If it is a threaded pipe, then some type of sealant should be placed on the threads before tightening.

You may also come across an occasional bronze fitting or valve. Brass is an alloy of copper and zinc, while bronze is an alloy of copper and tin. Other metals, such as antimony, may be used in alloys for various reasons. For amateur plumbing, there is probably little difference between the two. Brass will corrode more than bronze, which is why many valves are made of bronze rather than brass.

LEAD PIPE

Lead pipe is not used often anymore. Building codes allow it to be used for DWV and lead sleeves are still used on roofs to provide weather seals around vents. See Figure 2.11. However, lead is very heavy, soft, and rather expensive. It should never be used for a supply line because of the danger of lead poisoning.

ELECTRICAL GROUNDING

Another consideration in working with pipe is when the pipe is used for an electrical ground. The pipe may be used as a source for a ground often needed to complete an electrical circuit. Metallic pipes are often connected to the electrical distribution panel of the building as an electrical safety precaution. A ground clamp, shown in Figure 2.12, is typically used.

FIGURE 2.11   A lead roof vent sleeve for a DWV pipe.

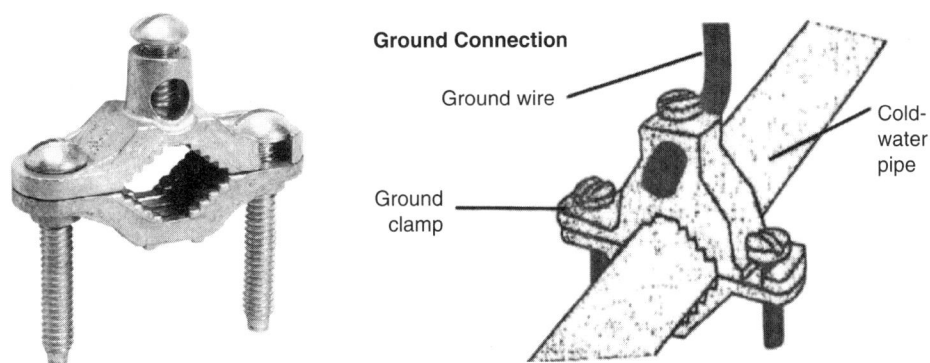

FIGURE 2.12 Ground clamp and one attached to a plumbing cold water line.

Metal pipes are buried, thereby making them a good ground for electrical equipment. To ground a pipe that isn't buried, a rod or a heavy wire is driven deep into the ground where the soil stays moist. Then a ground clamp is used to connect the pipe to the ground wire. This clamp should have a dielectric plate to prevent electrolytic action between two different metals.

The ground clamp can be used to ground a galvanized pipe to a copper ground wire. This is only needed in one location for a house, farm building, or an office building. There is one more grounding application. If, for example, you repair a leak in a copper pipe with a piece of plastic pipe, you must have disrupted the continuity of the grounded pipe. To maintain the integrity of this ground, use two clamps and a wire to connect the two metal sections that are separated by the plastic patch (refer to Figure 2.13).

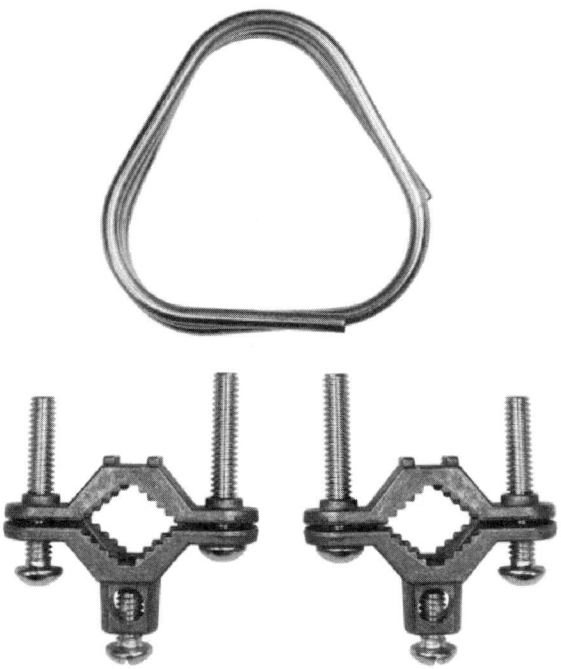

FIGURE 2.13 Copper wire and two grounding clamps to connect copper pipe over a PEX fitting plastic pipe repair to maintain continuity.

REVIEW QUESTIONS

1. Which type of pipe has been discontinued since 1992 due to health concerns?
   - a. galvanized steel
   - b. brass
   - c. copper
   - d. asbestos cement

2. Which of the following organizations has standards pertaining to pipe?
   - a. ANSI
   - b. ASTM
   - c. ASME
   - d. API
   - e. all of these
   - f. none of these

3. Which pipe grade requires the least yield strength requirement for pipe?
   - a. A
   - b. B
   - c. C
   - d. PSL 2

4. Which letter may be used to grade API pipe in lieu of A, B, and C?
   - a. S
   - b. R
   - c. P
   - d. X

5. In most cases, how are pipe and tubing measured?
   - a. tubing and pipe are always measured by their outside diameter
   - b. tubing and pipe are always measured by their inside diameter
   - c. tubing is measured by its outside diameter and pipe is measured by its inside diameter
   - d. tubing is measured by its outside diameter and pipe is measured by its NPS

6. Which of the following is correct?
   - a. STD and Schedule 20 are always the same
   - b. STD and Schedule 30 are always the same
   - c. STD and Schedule XS are always the same
   - d. XXS and XXH are always the same

7. What does NPS stand for?
   - a. National Pipe Standard
   - b. National Pipe Society
   - c. Nominal pipe size
   - d. Nominal pipe strength

8. The XH pipe weight designation is similar to?
    a. Schedule 20   b. Schedule 40
    c. Schedule 80   d. Schedule 160

9. Which of the following has a thicker wall up to 6-inches diameter?
    a. STD   b. XS
    c. XH   d. XXH

10. What could be the designation for pipe that has one beveled end?
    a. BE   b. BOE
    c. TOE   d. BVO

11. What type of applications do low-pressure pipe classifications refer to?
    a. oil/gas pipeline   b. steam/boiler
    c. compressed air   d. plumbing

12. Which type of copper pipe has the thinnest walls?
    a. K   b. L
    c. DWV   d. M

13. Which type of copper pipe is identified with green lettering?
    a. K   b. L
    c. DWV   d. M

14. What does ACR/MED refer to?
    a. annealed C rated medium pipe
    b. annealed C rated medical use pipe
    c. air conditioning and rated and for medical use
    d. air conditioning and refrigeration and for medical use

15. Which type of copper pipe does the size refer to its outside diameter?
    a. ACR/MED   b. K
    c. L   d. DWV

16. Why do metal pipes make a good ground for electrical equipment?

   a.   they are buried   b.   they conduct electricity

   c.   they are durable   d.   all of these

17. How do you ground a pipe that is not buried?

   a.   use a rubber strap

   b.   connect to a PVC drain pipe

   c.   hammer a long rod or heavy wire into the ground and connect to the pipe

   d.   none of these

18. What does a ground clamp require for?

   a.   photovoltaic plate   b.   diaconate plate

   c.   dielectric plate   d.   dielectric membrane

19. What is typically used to ground a galvanized pipe to a copper ground wire?

   a.   solder   b.   flange

   c.   ground clamp   d.   grounding wire nut

20. What must be done when using plastic pipe to repair a copper pipe?

   a.   nothing

   b.   attach a grounding clamp below the repair to another one above the repair with a wire

   c.   apply metal tape on the plastic pipe

   d.   cannot repair a copper pipe with a plastic pipe

ANSWERS TO REVIEW QUESTIONS

1. d	2. e	3. a	4. d	5. d
6. d	7. c	8. c	9. d	10. b
11. d	12. c	13. a	14. d	15. a
16. d	17. c	18. c	19. c	20. b

Chapter 3

PLASTIC PIPE STANDARDS AND IDENTIFICATION

Performance Objectives

After studying this chapter you will (be able to):

1. Identify various types of plastic pipe.

2. Understand how to join pieces of plastic pipe.

3. Determine how to handle plastic pipe.

4. Distinguish where plastic piping can be used.

5. Know how to bend plastic pipe.

PLASTIC PIPE

There are two types of pipe to work with as pipefitters: plastic and metal. Of these two types there are sub-divisions or specific metals and the same is applicable for plastic. This chapter examines the plastic types used every day in both domestic and commercial applications. It should be noted that plastic drain, waste, and vent (DWV) piping has been approved by local and state codes including the Building Officials Conference of America, Southern Building Code Congress, International Association of Plumbing and Mechanical Officials, and the Federal Housing Administration (FHA). Plastic pressure piping for hot and cold water supply is now permitted in FHA-financed rehabilitation projects. Plastic pipe enjoys markets in natural gas distribution, rural potable water systems, crop irrigation, and chemical processing. Almost 100 percent of all mobile homes and travel trailers have plastic pipe. Plastic pipe sales have skyrocketed over the years and there are now quite a few manufacturers in the United States as illustrated in Table 3.1. Some of the plastic types of pipes made in the United States are discussed in subsequent sections.

Polyvinyl Chloride

Polyvinyl chloride (PVC) has excellent properties, especially in chemical applications. PVC pipe is resistant to corrosion and chemical attack from most acids, alkalis, salts, fungi, and bacteria. Schedule 80 pipe can be threaded, but Schedule 40 should not be threaded. Schedule 40 pipe is dual marked as DWV and is white in color while Schedule 80 is typically dark gray in color. PVC is available not only in pipe, but also in complete and standardized lines of threaded and socket-type fittings as well as threaded and socket-type flanges. See Figure 3.1. Valves and pumps are also available. For corrosive atmospheres even PVC bolts and nuts are available for flanged connections.

Joining Methods

There are three recommended methods of joining PVC pipe: solvent welding, threading, and hot welding.

1. **Solvent Welding.** This is accomplished with primer, solvent cement, and an applicator. After the pipe end is cut square, using a hand saw and miter box or a power saw, clean the burs off with a deburring tool or knife and liberally brush PVC primer inside the fitting or pipe and the other PVC receiving pipe (typically twice). While the primer is still wet, liberally apply the PVC cement over the primered PVC parts (preferably twice). Pipe should then be pressed firmly into the fitting and turned a quarter of a turn for even distribution of cement. Handling is allowed after 30 seconds. Full strength of the solvent weld is not reached for 24 or more hours depending upon the temperature. See Figure 3.2.

TABLE 3.1   Number of pipe manufacturing facilities by state.

Alabama	9	New York	10
Alaska	1	North Carolina	22
Arizona	9	North Dakota	2
Arkansas	7	Ohio	27
California	40	Oklahoma	9
Colorado	9	Oregon	6
Connecticut	2	Pennsylvania	19
Florida	27	Puerto Rico	4
Georgia	12	Rhode Island	1
Idaho	4	South Carolina	4
Illinois	10	Tennessee	10
Iowa	10	Texas	55
Indiana	9	Utah	4
Kansas	7	Vermont	2
Kentucky	9	Virginia	2
Louisiana	9	Washington	9
Maryland	2	West Virginia	2
Massachusetts	8	Wisconsin	5
Michigan	4	Wyoming	3
Mississippi	5	**Total**	**445**
Missouri	11		
Montana	2		
Nebraska	8		
Nevada	8		
New Hampshire	1		
New Jersey	5		
New Mexico	4		

2. **Threading.** This method is not recommended for schedules A and 40 or in any case where operating temperature will exceed 120°F. Using standard hand or machine threading tools with sharp dies, schedules 80 and 120 can be easily threaded without the use of cutting lubricants. Soft pads should be placed in pipe vise jaws to prevent scaring of pipe. A tapered plug should be inserted in the pipe to ensure thread uniformity. In the threaded assembly, screwed fittings should be started carefully, and hand-tightened, then further tightened with a strap wrench because standard pipe wrenches cannot be used since they can deform and scar the pipe, thus weakening or even breaking it.

3. **Hot Air Welding.** Hot air welding or air fusion welding of PVC pipe can be done effectively after instruction and practice in using the technique. When properly made, hot welds have average tensile strength of 80% to 90% of the PVC material itself. Hot air welding equipment and filler rods that are designed for PVC welding are available from several manufacturers.

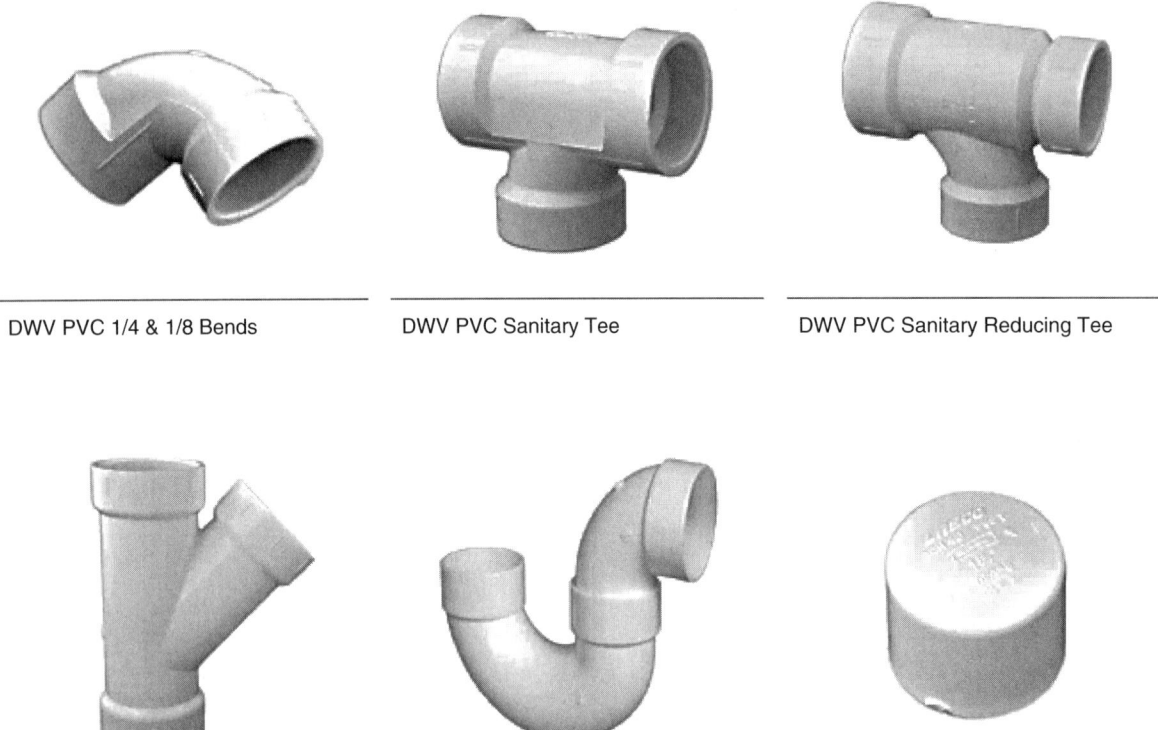

| DWV PVC 1/4 & 1/8 Bends | DWV PVC Sanitary Tee | DWV PVC Sanitary Reducing Tee |
| DWV PVC Wye | DWV PVC P Trap | White PVC Socket Caps |

FIGURE 3.1   A variety of PVC fittings.

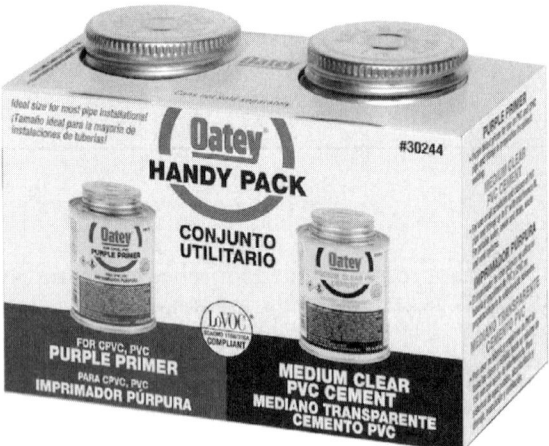

FIGURE 3.2   PVC primer and solvent cement (Oatey).

Pipe Supporting

Roll, ring, angle, or spring hangers may be used to support PVC pipe. Most favorable is a clevis or saddle-type hanger, except where thrust along the axis of the pipe must be controlled as in thermal expansion. For firm anchoring, metal compression hangers are satisfactory only when padded with a compressible insert. Pipe lines should have additional support to that recommended in the table of fittings and flanges.

TABLE 3.2   Recommended PVC line support distances in feet so pipe will not bend and break.

Operating Temperature, Deg. F	Recommended Support Spacing, in feet, for Polyvinyl Chloride Pipe					
	Nominal Pipe Size, inches					
	½–¾	1–1¼	1½–2	3	4	6
Schedule 40						
60	5.5	6.1	6.5	7.7	8.0	8.7
100	4.8	5.4	5.7	6.8	7.1	7.7
120	4.2	4.5	4.8	5.8	6.0	6.4
130	3.7	4.0	4.3	5.1	5.4	5.8
140	3.2	3.5	3.7	4.5	4.7	5.0
Schedule 80						
60	6.5	7.3	7.7	9.0	9.6	10.8
100	5.7	6.4	6.8	8.0	8.4	9.5
120	4.8	5.5	5.8	6.8	7.1	8.0
130	4.3	4.8	5.1	6.0	6.4	7.2
140	3.7	4.2	4.5	5.2	5.6	6.3
Schedule 120						
60	6.8	7.7	8.2	9.8	11.0	12.3
100	6.0	6.8	7.2	8.6	9.6	10.9
120	5.0	5.8	6.1	7.2	8.1	9.2
130	4.5	5.1	5.5	6.5	7.3	8.2
140	3.9	4.5	4.8	5.6	6.3	7.1

Note: Spacings apply to uninsulated lines carvine fluids

Refer to Table 3.2. Valves should be supported independently, and braced to resist twisting during opening and closing. Continuous supports are advisable for pipe carrying hazardous fluids and for lines operating at the upper thermal limits of the PVC pipe: 1400 for type I—normal impact and 1500 for type II—high impact.

Buried Lines

For all buried lines, cemented (solvent welded) joints are recommended. An additional one-bead seal weld (hot air welded) will give added assurance of a permanent, leak-proof joint. When possible, the pipeline should be assembled, wholly or in sections, above ground and then lowered into a prepared trench. Then fill the bottom, 4 to 6 inches below the pipe. It should be free of rocks and other sharp objects. The same type of fill is recommended for the first 8 to 12 inches and backfill should fully enclose the line. All lines should be laid below the frost level. Some manufacturers advise running cool water through the pipe to shrink it to normal length before and during the process, especially in hot weather.

Thermal Expansion

PVC like all thermoplastic pipe materials has a relatively high rate of expansion to temperature change as compared to ferrous materials. As with most metals the operating temperature range will dictate the type of joint and the means of providing for expansion. Thermal compression of a solvent welded PVC pipe line is not required when operating extremes do not differ from installation temperatures by more than plus or minus 30°F. However, solvent welded or hot air welded connections are preferable to threaded joints where any appreciable temperature changes are expected. In both cases, axial restraint guides for the thrust along the axis of the pipe are required. In order to compensate for thermal expansion and contraction, any commercial expansion joints may be used. In addition, U-bend offsets made from hot-formed PVC pipe or from PVC fittings and straight lengths of pipe are suitable. If enough space is available, this latter method is the most advisable since it offers the advantages of a continuous PVC pipeline with moving parts and no change in quality of corrosion resistance. The PVC line should be installed using the same expansion offset practices as followed for a steel line expanding to the extent, in inches per 100 feet.

Chlorinated Polyvinyl Chloride

Chlorinated polyvinyl chloride (CVPC) is a thermoplastic produced by the chlorination of PVC resin, which is significantly more flexible and can withstand higher temperatures than standard PVC. Uses for CPVC include hot and cold water delivery pipes and industrial liquid handling. CPVC is more expensive so it is typically only used for hot water lines that require the extra heat resistance. It can handle 200°F when standard PVC can only handle temperatures of up to 140°F safely. CPVC can be joined in the same fashion as PVC; however, it does take a different solvent cement and primer for proper adhesion. CPVC Schedule 40 is typically tan in color and Schedule 80 is typically light gray in color. Some manufacturers use either color so make sure to read the label printed on it before use.

Acrylonitrile-Butadiene-Styrene

The three most commonly used plastic pipe (not flexible tubing) are PVC, CPVC, and acrylonitrile-butadiene-styrene (ABS). ABS is only used for DWV. It requires fittings made of ABS and special ABS cement. However, the same tools may be used for both PVC and CPVC. When cementing plastic pipe in place, the cement is actually a solvent and the process is really a form of welding. The procedure to fasten ABS pipe is as follows: First, measure each piece of pipe. Figure 3.3 shows the allowances that should be made for the joints and fittings.

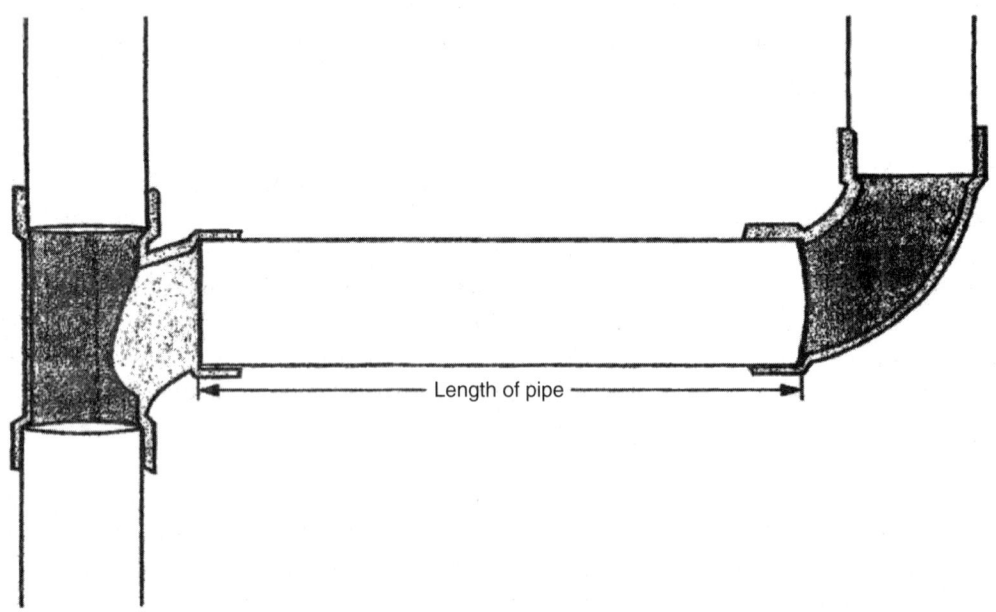

FIGURE 3.3   Fitting allowances for each end of the pipe.

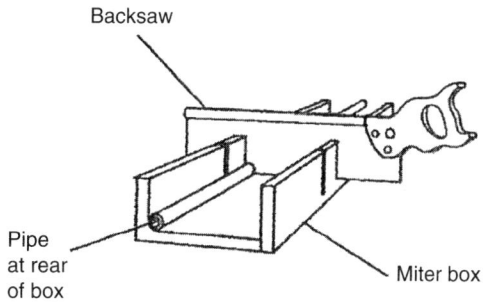

FIGURE 3.4 Back saw and miter box used to cut plastic pipe.

Next, cut the plastic pipe with a tubing cutter or a common woodcutting saw as shown in Figure 3.4.

To join the pipe together, simply coat the inside of the fitting and the outside of the pipe with cement. See Figure 3.5. Do this with the built-in swab that is connected to the lid of the can. Remember, no primer is needed with ABS plastic. Then push the two pieces together with a slight twisting motion, as shown in a step in Figure 3.5.

ABS pipe can be joined in the same three ways that PVC and CPVC pipe are joined; however, all three take different solvent cements and filler rods. It should be noted that ABS provides the lightest weight of all semi-rigid or rigid thermoplastic pipes. ABS is supplied in 20-foot lengths in ½- to 6-inch diameters and is available in schedules 40, 80, and 120. See Figure 3.1. A variety of threaded and socket-type fittings and flanges, and valves are provided in ABS to give the pipe a wide range of application within its limits of chemical resistance. ABS is black in color. The advantage of ABS pipe over PVC is that only a solvent needs to be applied before attaching the pipes together. In addition, less time is required to hold them together. ABS is also stronger and works well for buried DWV pipe and can withstand colder temperatures. However, ABS tends to warp and bend when directly exposed to sunlight. Check the building codes in an area before using ABS or PVC pipe.

ABS pipe and fittings were originally developed in early 1950s for use in oil fields and the chemical industry. In 1959, John F. Long, a prominent builder from Arizona, used ABS pipe in an experimental residence. Twenty-five years later, an independent research firm dug up the pipe and analyzed a section of the drain pipe. The result: no evidence of rot, rust, or corrosion. ASTM standard for ABS-DWV pipe and fittings was originally approved in 1967.

There are many advantages to the use of ABS piping in new construction. First, it can be relied upon to serve for more than 50 years without signs of rot, rust, or corrosion. Some of the advantages of ABS are as follows:

- ABS is easier and less expensive to install than metal pipe.

- It has a superior flow due to its smooth interior finish.

- Does not rot, rust, or corrode and does not collect waste.

- This type of piping can withstand earth loads and shipping (with proper handling).

- ABS pipe resists mechanical damage, even at very low temperatures.

- It can perform at an operational temperature up to 140°F (60°C).

- The pipe is lightweight (one person can load and unload it).

- It takes less time to rough-in than metal or PVC DWV lines.

- Dimensions and pressure limits for ABS pipe are found in Table 3.3.

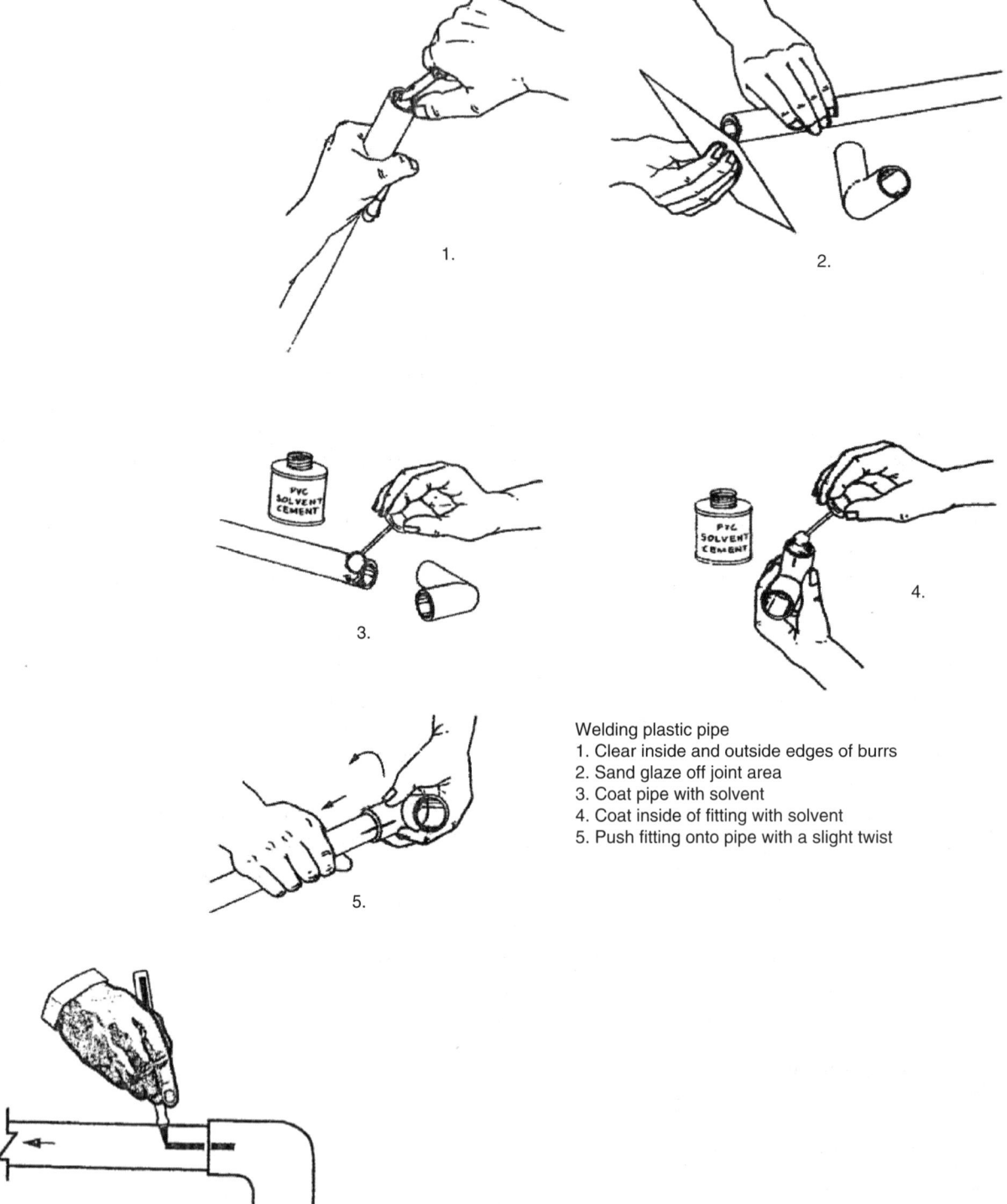

Welding plastic pipe
1. Clear inside and outside edges of burrs
2. Sand glaze off joint area
3. Coat pipe with solvent
4. Coat inside of fitting with solvent
5. Push fitting onto pipe with a slight twist

FIGURE 3.5   Attaching an ABS fitting to an ABS pipe with solvent cement.

TABLE 3.3 Pressure limits of various size ABS pipe.

Dimensions and Pressure Limits of ABS Pipes				
Nominal Size, inches	Outside Diameter, inches	Inside Diameter, inches	Max, Operating Pressure at 70 Deg. F, psi	Max Operating Pressure at 170 Deg. F, psi
Schedule 40				
½	0.840	0.622	150	75
¾	1.050	0.844	150	75
1	1.315	1.049	125	60
1¼	1.660	1.380	100	50
1½	1.900	1.610	90	45
2	2.375	2.067	75	40
Schedule 80				
½	0.840	0.546	300	150
¾	1.050	0.742	300	150
1	1.315	0.957	250	125
1¼	1.660	1.278	200	100
1½	1.900	1.500	175	90
2	2.375	1.939	150	75

The disadvantage to ABS pipe is that it is susceptible to degrading if a lot of chlorine is used in a community's drinking water to kill bacteria. There seems to be fewer issues when used in colder climates than in the Southern United States.

WORKING WITH SOFT PLASTIC TUBING

To work with soft plastic tubing, such as polyethylene (PE) or polypropylene (PP) PE or PP tubing, special fittings known as barbed fittings need to be used. Refer to Figure 3.6. Space should be allowed for the fitting shoulder in flexible tubing because the shoulder is on the outside rather than the inside.

Plastic tubing is easy to cut with a special tubing cutter or with ordinary woodworking tools. PE tubing and PP tubing are not totally limp so they can be held in a pipe vise or the pipe jaws of a bench vise. A lot of force is not necessary to hold them. Therefore, only tighten the vise just enough to firmly hold the work.

The fittings can usually be pressed or pushed into the opening of the tubing, but it may take some muscle. Larger sizes can be very stubborn. The fitting may require a board and hammer, as seen in Figure 3.7. Be sure to tap the board gently. Allow the pipe to project from the vise or hand hold it the length that the fitting will take. Again, see Figure 3.7. A lot of force can distort the soft fitting and can result in a leak. Persistence and patience work better than strong-arm tactics that will deform the plastic and cause it to leak.

Cellulose Acetate Butyrate

Cellulose acetate butyrate (CAB) tubing is manufactured from a product of the chemical processing of cellulose acetate and butyric acid. It can be manufactured in clear form for observation of flow, or it can have a non-toxic black pigment added to provide even better ultraviolet (UV) ray resistance for outdoor

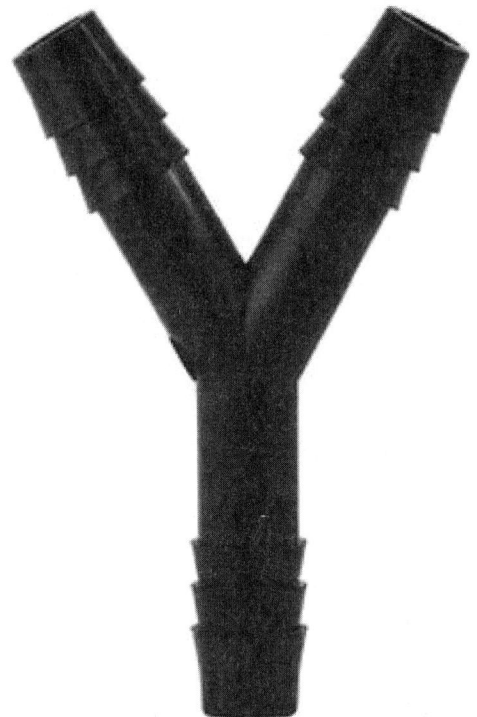

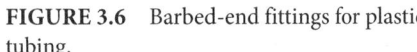

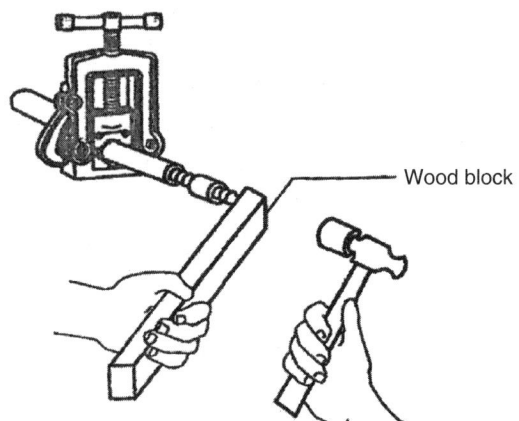

FIGURE 3.6   Barbed-end fittings for plastic tubing.

FIGURE 3.7   Gently tapping a barbed end fitting into a plastic pipe with a board and hammer.

installation. Ditching for underground installations does not require exacting dimensions with CAB. The semi-rigidity of this tubing permits its handling in straight lengths yet provides sufficient flexibility so that it can follow the contour of the ground. When making underground installations, it is recommended that the tubing be snaked in the ditch. For each 100 feet of butyrate tubing, a minimum of ⅛ inch per degree of temperature difference (that it will be subjected to) should be allowed for thermal expansion or contraction. Concrete or dirt backfill should be poured shortly after the tubing has been placed.

CAB, typically referred to as butyrate, is extruded in nominal diameters from ½ through 6 inches in both flexible and hard tubing, depending upon how much plasticizer has been added to it. CAB is impact- and UV-resistant with excellent wearing and weathering characteristics. CAB is unique tubing that can be bonded to fittings with solvent cements such as PVC pipe. Butyrate has characteristics of pipe and tubing and works better than either in certain applications.

APPLICATIONS OF PLASTIC PIPE

The major application of plastic pipe in general has been in jet-pump installations and other water system piping, including water services, sprinkling, and plumbing drainage. Other fields served well by thermoplastics according to their characteristics and limitations include food, beverage, and chemical processing for industry, air conditioning recirculating lines, swimming pools, skating rinks, and natural gas and oil field piping.

Radiant heating has been, and is, considered a borderline application of these thermoplastics. A number of systems utilizing plastic pipe have been installed for radiant heating, many have been successful and some have been troublesome. Manufacturer reluctance to positively recommend any thermoplastic for radiant heating is based not on lack of knowledge of capability of the particular material involved, but rather on the possibility that any overheating created in the system would cause softening and possible failure of the pipe.

RESISTANCE TO CHEMICAL CORROSION

Corrosion of metals is usually of the galvanic or electrochemical type, characterized by a minute flow of electrical current from anode to cathode areas. This is accompanied by a loss of metal to the surrounding environment. Consequently, corrosion rates can readily be measured by weight lost.

Corrosion takes place in a different manner with plastics. This is because plastics are nonconductive. Galvanic and electrochemical effects are nonexistent. Pitting and grooving do not take place, and there is no loss of material from the corrosion-causing fluid. Plastic corrosion is an absorption type of reaction in which the corrosive media actually penetrates or diffuses into the plastic. For this reason, corrosion is normally associated with weight gain rather than as weight loss.

Thermoplastic resistance to corrosion is influenced by concentration, temperature, and stress. The ability of the principal thermoplastic to handle various acids and other chemicals is better than most metals; however, most metals can withstand higher pressure ratings than plastics.

PEX TUBING

PEX tubing is made for plumbing houses and businesses. It has advantages in many ways. First, it is a tubing item that is made from a crosslinked high-density polyethylene (HDPE) polymer. The HDPE is melted and continuously extruded into a forming tube. The crosslinking of the HDPE is accomplished in one of three different methods.

PEX has been used for plumbing in the United States since the 1980s. It had, however, been used in Europe for about 10 years prior to its acceptance in the United States. The PEX system is very useful and used in a lot of the current residential construction. It is less expensive than metals such as copper pipe and much lighter and easier to work with. See Figure 3.8 for a crimping tool designed to work exclusively with PEX in plumbing jobs.

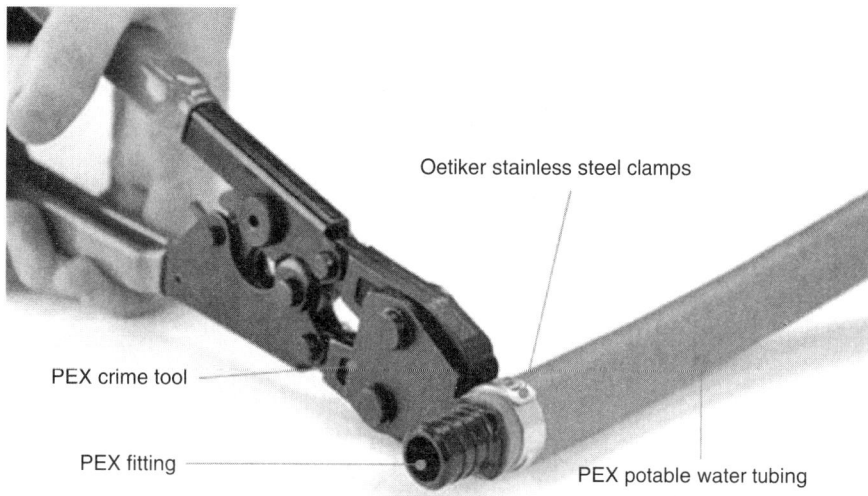

FIGURE 3.8 PEX stainless steel clamp (SSC) crimping tool.

Crosslinking

This is a chemical that occurs between polyethylene polymer chains. Crosslinking causes the HDPE to become stronger and resistant to cold cracking or brittleness on impact while retaining its flexibility. The three methods of crosslinking HDPE are the Engels method (PEX-a), the Silane method (PEX-b), and the Radiation method (PEX-c). Several industry participants claim that the (PEX-a) method yields more flexible tubing than the other methods. All three types of PEX tubing meet the ASTM, NSF/ANSI, and CSA standards. The terms PEX pipe and PEX tube are used interchangeably; however, some manufacturers distinguish between the two by manufacturing to different inside/outside diameters. For example, PEX pipe may be manufactured to CIS-OD (copper tubing size, outside diameter controlled) sizes, commonly with a standard thickness of standard dimension ratio (SDR).

Colored PEX

Before extrusion the HDPE can be pigmented to yield color-coded pipe. Common PEX tubing are "natural" (a hazy clear, unpigmented), white, black, red, and blue. The red and blue colors are used to help plumbers and homeowners distinguish between hot- and cold-water supply lines. The tube will be marked on the outside to show which standards it meets. Then, as it is produced, PEX is wound onto spools for storage and shipping. A typical spool of ½-inch PEX will hold 1200 feet of tubing.

PEX tubing is manufactured to CTS-OD sizes in the most common sizes including ⅜-, ½-, ⅝-, ¾-, and 1 inch. The standard method of connecting PEX to brass PEX uses a copper crimp ring and a PEX crimping tool. See Figure 3.9.

PEX Connecting Methods

There are three methods of connecting PEX, one of which is the expansion method. This method was developed as a proprietary solution and is currently available from one company. Information about testing standards for this method can be found in the ASTM standards. The expansion method involves using an expansion tool to increase the diameter of the PEX tube. This shrinks back to shape around the fitting. A plastic ring is then pressed over the fitting to ensure a tight fit and connection.

Another method of connecting PEX tubing is referred to as the stainless steel clamp (SSC). The fittings used here are the same used in the standard connection method. However in this method, the SSC fastens

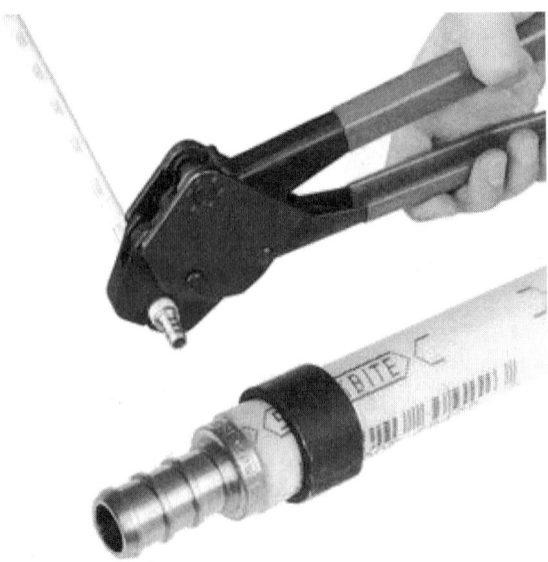

FIGURE 3.9 PEX copper crimp ring tool.

the PEX tube and fitting. A special SSC crimping tool is used to the clamp around the tube and fitting. More details about this method can be found in the ASTM standards. The SSC method uses special clamps designed for PEX connections. The fittings used here are the same as used in the standard connection method. However, in this method the SSC fastens the PEX tube to the fitting. A special SSC crimping tool is then used to tighten the clamp around the tube fitting. More information on this method can be found in the ASTM standards and on the Internet.

The third method for making connections between PEX tubing is the compression method. For moderate-to large-size jobs this method is more expensive than using the standard connection method since compression fittings cost more than PEX fittings. Many companies make fittings to connect PEX to PEX or PEX to copper, PVC, and/or other materials as well. These fittings use one or more of several threaded compression nuts. These fittings are faster and easier than most competing methods, but cost more per fitting than standard PEX fittings. Most of the fittings are made of bronze or copper. Some manufacturers make engineered plastic fittings for PEX. These engineered plastic fittings have ridges (see Figure 3.9). The characteristic ridges on the inset part of the fitting distinguish a PEX fitting from other fittings. The ridges, the flex tube, and the crimped copper ring all work together to form a high-pressure seal.

Tools for Working with PEX

Three basic tools are needed to work with PEX tubing. The main ones are the pipe cutter, crimping tool, and a de-crimping tool. The pipe cutter is used to make a clean, square cut before inserting the tubing into the fitting. The crimping tool squeezes or compresses the one piece tightly over the other piece. The main crimping tool can be purchased in several configurations from various venders. One popular model has the capability of crimping either ½- or ¾-inch PEX tube. Other sizes are also available. A de-crimping tool is designed to remove the copper crimp ring from the tube and fitting. See Figure 3.10.

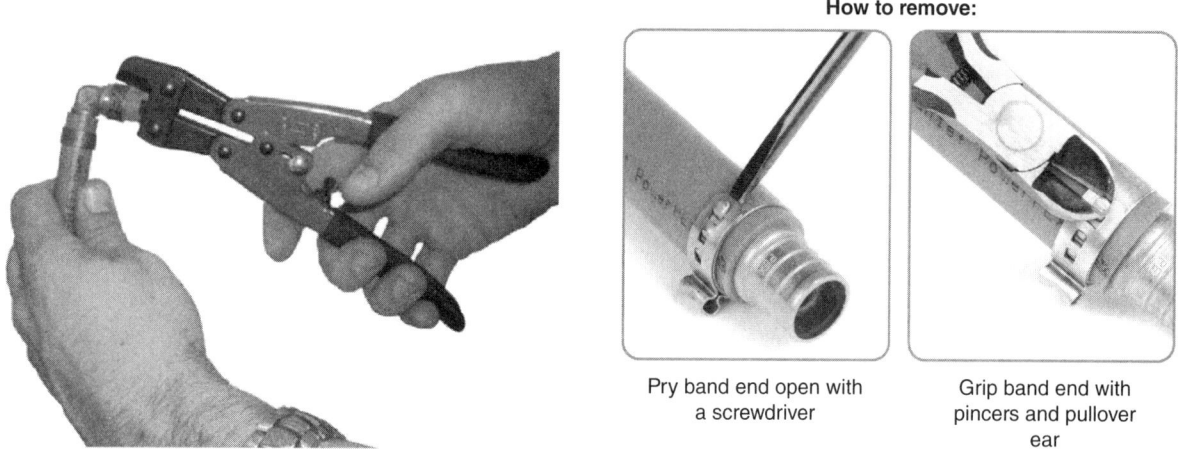

How to remove:

Pry band end open with a screwdriver

Grip band end with pincers and pullover ear

FIGURE 3.10 Pictured left, a de-crimping tool required for copper crimp rings and right, most stainless steel clamps can be removed by prying off with a screwdriver or cutting off the tab with diagonal cutters.

REVIEW QUESTIONS

1. Which of the following are the three recommended ways of joining PVC pipe?
   a. solvent cement, threading, and bolting
   b. solvent cement, threading, and ultrasonic welding
   c. solvent cement, threading, and hot air welding
   d. solvent cement, threading, and friction welding

2. How many times should primer be applied to PVC pipe before cementing?
   a. once
   b. twice
   c. never
   d. only apply if the pipe is dirty

3. How must PVC pipe be fitted to ensure even distribution of cement?
   a. firmly with a quarter turn
   b. firmly with a half turn
   c. pressed until it bottoms out in the fitting
   d. none of these

4. How long does it take PVC cement to reach its full strength?
   a. 30 seconds
   b. 30 minutes
   c. 12 hours
   d. 24 hours

5. What schedule PVC pipe can be threaded?
   a. 40
   b. 60
   c. 80
   d. none of these

6. What type of wrench should be used to tighten threaded PVC?
   a. open end
   b. pipe
   c. spanner
   d. strap

7. What is the tensile strength of hot air welded PVC joints?
   a. 25%
   b. 50%
   c. 75%
   d. 80%

8. Which form of PVC is placed in a joint when hot air welded?

   a. powder   b. liquid

   c. spray   d. filler rod

9. Which type of plastic pipe is only used for DWV?

   a. ABS   b. PVC

   c. CPVC   d. PEX

10. Joining PVC parts together with a solvent is really a form of what?

    a. adhesion   b. welding

    c. cementing   d. none of these

11. How much primer should be used when attaching ABS pipe?

    a. one coat is all that is needed

    b. two coats are all that is required

    c. two coats need to be applied to the pipe and fixture

    d. none of these

12. Which type of fittings is used primarily on plastic tubing?

    a. needle   b. barbed

    c. clamp   d. pressure

13. How are fitting applied to plastic tubing?

    a. pressed or pushed on   b. with a tubing puller

    c. with a pair of pliers   d. none of these

14. Which type of PVC pipe is rated high impact?

    a. Type I   b. Type II

    c. Type III   d. HI grade

15. What is the recommended method for connecting PVC lines buried underground?

    a. tape   b. cement

    c. hot air welded   d. threaded

16. What schedule is ABS pipe available in?

    a. 20, 40, 60   b. 40, 60, 80

    c. 40 and 60   d. 40 and 80

17. In most cases, what color is ABS pipe?

    a. black   b. white

    c. tan   d. gray

18. Which plastic pipe is the strongest in colder temperatures and works best for buried DWV?

    a. ABS   b. PVC

    c. CPVC   d. CAB

19. Which type of plastic can be molded in clear?

    a. ABS   b. PVC

    c. CPVC   d. CAB

20. Which of the following is NOT a recommended use of plastic pipe?

    a. AC recirculating lines   b. swimming pool lines

    c. oil field lines   d. radiant heating lines

21. What are the three most common types of plastic pipe?

    a. PEX, PVC, ABS   b. PEX, PVC, CPVC

    c. ABS, PVC, CPVC   d. CAB, PVC, ABS

22. What is butyrate pipe manufactured from?

    a. cellulose acetate and butyric acid

    b. cellulose acetate and butadiene

    c. polybutylene

    d. none of these

23. Approximately, how many plastic pipe companies are there in Texas?

    a. 10   b. 20

    c. 50   d. 100

24. What size of butyrate tubing is available?

    a. ⅛–12 inches

    b. 1–12 inches

    c. ½–6 inches

    d. ½–20 inches

25. What type of plastic is concerned with crosslinking?

    a. ABS

    b. PVC

    c. CAB

    d. PEX

ANSWERS TO REVIEW QUESTIONS

1. c	2. b	3. a	4. d	5. c
6. d	7. d	8. d	9. c	10. b
11. d	12. b	13. a	14. b	15. b
16. d	17. a	18. a	19. d	20. d
21. c	22. a	23. c	24. c	25. d

—NOTES—

Chapter 4
SOLDERING AND BRAZING NONFERROUS PIPE AND TUBING

Performance Objectives

After studying this chapter you will (be able to):

1. Determine the difference between soft solder and hard solder.

2. Discuss the purpose of flux in soldering.

3. Understand how flux works.

4. Define the term spelter.

5. Know the distinction between melting points.

6. Discuss what adding zinc to solder does.

7. Know that solder is an alloy of lead.

8. Identify what sal-ammoniac is and where it is used and when.

9. Learn what is the most common solder alloy mix.

10. Understand when lead is used when soldering water supply pipes.

11. Determine where rosin-core solder is used.

12. Recognize when you would use acid core solder.

SOLDERING AND BRAZING

Solders for joining metallic surfaces or edges are almost always composed of an alloy of two or more metals. The solder used must have a lower melting point than the metals to be joined by it, but the fusing point should approach, as nearly as possible, that of the metals to be joined so that a more tenacious joint is the result. Solders may be divided into two general classes: hard and soft. The former fuses at a red heat, the latter, at a comparatively low temperature. See Figure 4.1. Moreover, soft soldering is conducted under 840°F (450°C) while hard soldering and brazing requires a temperature above that.

Soft Solders

Soft solders consist chiefly of lead and tin, although other materials are occasionally added to lower the melting point. Lead-tin alloys melt at a lower temperature with an increase in the percentage of tin, up to a certain point. However, when the tin exceeds 67 per cent, the melting point rises gradually to the melting point of tin. Soft solders are termed "common," "medium," and "fine." According to the tin content, those containing the most lead are the cheapest and have the highest melting temperatures.

Fine solders are largely used for soldering Britannia metals: such as brass- and tin-plated articles. It is also used for soldering cast iron, steel, copper, and many alloys. The soft solder called "common" is used by plumbers for ordinary work; this solder contains two parts of lead to one part of tin. The best soft solders are made from pure lead and pure tin. Antimony is an objectionable impurity as it renders the solder less fluid when melted and tends to affect perfect adhesion of the surfaces.

Zinc also has an adverse effect on soft solder, causing it to flow sluggishly. Aluminum acts in a similar way. A small percentage of phosphorus renders soft solder very "lively" which means the solder has a tendency

HARD SOLDERING
SILVER SOLDER WIRE

SOFT SOLDERING
LEADFREE SOLDER WIRE

FIGURE 4.1 Hard soldering wire rods and soft soldering flexible wire. (Stannol)

to run freely. Adding too much phosphorous is also detrimental, and if added to thin the solder it should be in the form of phosphor-tin.

Hard Soldering and Brazing

Hard solder is used for joining such metals as copper, silver, gold, and alloys such as brass, German silver, and gun metal which require a strong joint and often a solder the color of which is near that of the metal to be joined. The hard soldering of copper, iron, and brass is generally known as brazing, and the solder is known as spelter. The operations of hard soldering and brazing are identified, and the two terms are often used interchangeably. According to common usage; however, there is the following distinction. Brazing is generally understood to mean the joining of metals by a film of brass, whereas hard soldering (which is the term used by jewelers) ordinarily means that "silver solder" is used as the uniting medium. For hard soldering or brazing, a red heat is necessary and borax is used as a flux to protect the metal from oxidation and to dissolve the oxides that have formed. Heating can in some instances be done with a soldering iron, but should be done with a blowpipe, blowtorch, gas forge, or a coke fire.

As a greater degree of heat is required to melt spelter than soft solder, brazed work will withstand more heat without braking or weakening parts which are soldered. The chief advantage of a brazed joint, however, lies in its superior strength. Before work is assembled for brazing, it should be carefully cleaned. The parts are then fastened together in the position they are to occupy when joined.

Fluxes for Soldering

As two pieces to be soldered must be thoroughly alloyed with the metal used as a solder, the temperature must be raised and maintained at such a point that inter-penetration can take place completely. It is necessary that the surfaces to be joined be perfectly clean and means must be provided to prevent oxidation during soldering because oxides tend to prevent interfusion. This is accomplished by using a coating of some substance that melts at the fusing temperature of the solder, and thus excludes the air.

ALLOYS FOR BRAZING SOLDERS

The alloys for spelters used for brazing are composed of copper-zinc alloys. The melting point of these alloys depends upon the percentage of zinc. As the proportion of zinc increases, the melting point is lowered. The fusing point of the spelter should be as close as possible to that of the article to be brazed, as a more resolute joint is thereby secured. An easily fusible spelter may be made of two parts zinc to one part copper, but the joint will be weaker when an alloy more difficult to fuse is employed. A spelter that is readily fused may be made of 44% copper, 50% zinc, 4% tin, and 2% lead. Alloys containing much lead should be avoided, since lead does not transfuse with brass and thus decreases the strength of the joint. A hard solder for the richer alloys of copper and zinc may be produced from 53 parts copper and 47 parts zinc. Copper and iron have a much higher melting point than brass, thus allowing the use of a richer copper alloy.

Sweating

When parts are soldered together by heating them sufficiently to melt the solder, instead of using a soldering iron, the operation is often known as sweating. The finished surfaces forming the joint are first tinned or covered with solder. This is done by heating enough to melt the solder, then applying a flux (such as sal-ammoniac), and finally the solder.

There are some things you should know about solder. Solder is an alloy of lead and tin. Solders are rated by a ratio with the percentage of lead being the first number. For example, 60/40 solder is 60% lead and 40% tin. Because tin is more expensive than lead, 60/40 solder is cheaper. However, 60/40 solder may be more difficult to work with. Probably the best solder to use for copper pipe is 50/50 or 40/60 solder. The 50/50 solder is by far the more common of the two.

The most common shape of solder is wire solder. It also comes in bars and as paste. The paste is ready to use, but the bar solder is more difficult to use than wire solder.

Paste solder consists of tiny bits of solder suspended in a thick solution of flux. You simply brush it on, assemble the joint, and then heat it until the solder melts and fuses. You can see this happen around the edges.

Wire solder is the most common shape for the beginner. It is available in solid, acid-core, and rosin-core forms. The core type solders have a flux embedded in the center of the wire. The acid-core flux in plumbing solder varies from electrical solder, which contains a rosin-core flux. The acid in plumbing solder is very corrosive in order to remove any oxidation off the surface of pipes as the solder melts, thereby allowing it to adhere and form a waterproof joint.

Lead in the solder is a possible contaminant for water. Over a long period of time, the effects of the lead could cause problems for a person. To offset this potential hazard, some codes now require the use of lead-free solder. This type of solder is an alloy of tin and several other metals and works just as well as regular solder does. It doesn't contain lead and is a bit more expensive.

For joining copper or brass pipe, a soft solder (discussed in earlier sections) can be used. Silver solder has some disadvantages. It requires more heat, takes more time to work, and is more expensive than the others. Figure 4.2 has more information on actually working with the various types of silver solders. It should be noted that silver (Ag) has a higher melting temperature, so the higher the content of silver then the higher the temperature that is required to solder.

Name	Ag	Cu	Zn	Cd	Melt. Point	
					°C	°F
"IT"	80	16	4		809	1490
Hard	76	21	3		773	1425
Medium	70	20	10		747	1390
Easy	60	25	15		711	1325
"Easy Flo"	50	15	15	20	681	1270

FIGURE 4.2 Chemical composition of silver solders. IT stands for intense temperature and requires more heat than the other solders.

REVIEW QUESTIONS

1. What is the composition of most solders?
   a. two or more metals
   b. metal and a gas
   c. aluminum and another metal
   d. iron and another metal

2. What are the two main categories for solder?
   a. low and high
   b. lead and zinc
   c. soft and hard
   d. spelter and brass

3. What type of solder is used by plumbers?
   a. plumbers
   b. basic
   c. common
   d. fine

4. What temperature is required to do hard soldering?
   a. 400°F
   b. 550°F
   c. below 840°F
   d. above 840°F

5. What is primarily used for hard soldering?
   a. antimony
   b. spelter
   c. lead
   d. tin

6. Which is the strongest joint?
   a. soft soldered
   b. brazed
   c. borax
   d. tinned

7. What helps to lower the melting temperature?
   a. aluminum
   b. antimony
   c. copper
   d. zinc

8. What is typically the joining material in brazing?
   a. aluminum
   b. brass
   c. silver solder
   d. soft solder

9. What is typically the joining material in hard soldering?

    a. aluminum   b. brass

    c. silver solder   d. soft solder

10. What is the main joining material composed of for brazing?

    a. copper and lead   b. copper and tin

    c. copper and zinc   d. copper and antimony

11. Which type of solder is used primarily for plumbing?

    a. acid core   b. rosin core

    c. ribbon core   d. copper core

12. What is usually embedded in solder?

    a. sanitizer   b. ionized bath

    c. copper sulfate   d. flux

13. What type of solder helps clean copper pipes?

    a. acid core   b. borax core

    c. copper core   d. antimony core

14. Why is plumbing solder now lead-free?

    a. lead is too expensive

    b. lead is a contaminant for water

    c. lead did not work as well

    d. none of these

15. What form does solder mainly come in for brazing?

    a. powder   b. wire

    c. filler rod   d. none of these

ANSWERS TO REVIEW QUESTIONS

1. a	2. c	3. c	4. d	5. b
6. b	7. d	8. b	9. c	10. c
11. a	12. d	13. a	14. b	15. c

Chapter 5

WORKING WITH IRON, CLAY, AND FIBROUS PIPE

Performance Objectives

After studying this chapter you will (be able to):

1. List the steps involved with threading pipe.

2. Determine how to properly handle a die for threading pipe.

3. Discuss how to work with cast iron pipe.

4. Know how to cut cast iron pipe.

5. Understand what a tamping iron does.

6. Know what a flange is.

7. Identify vitrified clay pipe.

8. Discuss how to work with bituminous pipe.

9. Know how to work with galvanized steel pipe.

10. Determine how to repair a leaking fitting.

WORKING WITH CAST IRON PIPE

Working with cast iron pipe may require some special operations and tools. Always wear the appropriate safety gear when working with molten lead, an open fire, and hot metals.

Cast iron pipe is heavy and hard to work with. Perhaps the most exacting job in working with cast iron pipe is its cutting. It must be measured carefully since it is expensive and comes in long lengths. Also remember to measure twice and cut once. Make a visible mark all the way around the pipe where it is to be cut. Furthermore, make sure the proper allowance is given so the pipe will extend into the fitting.

A hacksaw can be used for cutting the pipe, but that's a long and tedious job. It is recommended that either a reciprocal saw or a chain pipe cutter saw be used. Both can be rented. A chain cutter as shown in Figure 5.1 is wrapped around the pipe, attached, tightened, and then turned just as one would do with a smaller pipe cutter. In order to cut the pipe, turn and tighten the cutter until the pipe comes apart or cut about halfway through the pipe and break off the end.

If there is another person to assist, a circular saw can be used with a metal cutting blade. One person holds the saw and makes a shallow cut on the mark while the other person steadies the pipe and rotates it slightly as it is cut. Refer to Figure 5.2.

Be very careful because the saw can slip, the blade can break, and there will be a lot of hot sparks. Eye protection, gloves, and natural fiber clothing are essential.

The old way to cut cast iron pipe was to score the intended cut with a hacksaw to a depth of about 1/16 to 1/8 inch. This is shown in Figure 5.3. Then a cold chisel is used to deepen and widen the cut. Sometimes the end piece fell off at some point during the chisel operation. If it didn't then a large hammer was used to give the waste end a series of smart raps all the way around the pipe until the end broke off. This way still works, but it is a lot easier with power tools.

Once the pieces are cut to length, they must be joined. There are some things to consider before doing this. If installing a main soil stack which is a vertical assembly, local codes will probably require that each vertical joint be sealed with oakum (loosely twisted rope) and lead. Whole lengths will have to be worked where hub joints are required, see Figure 5.4. The hub of the top pipe fits into the opening of the bottom pipe. Oakum is fitted into the opening of the bottom pipe. It is then fitted into the joint and firmly tamped (packed down) in place. Leave about ¼ to ½ inch between the packed oakum and the top of the flange. Next,

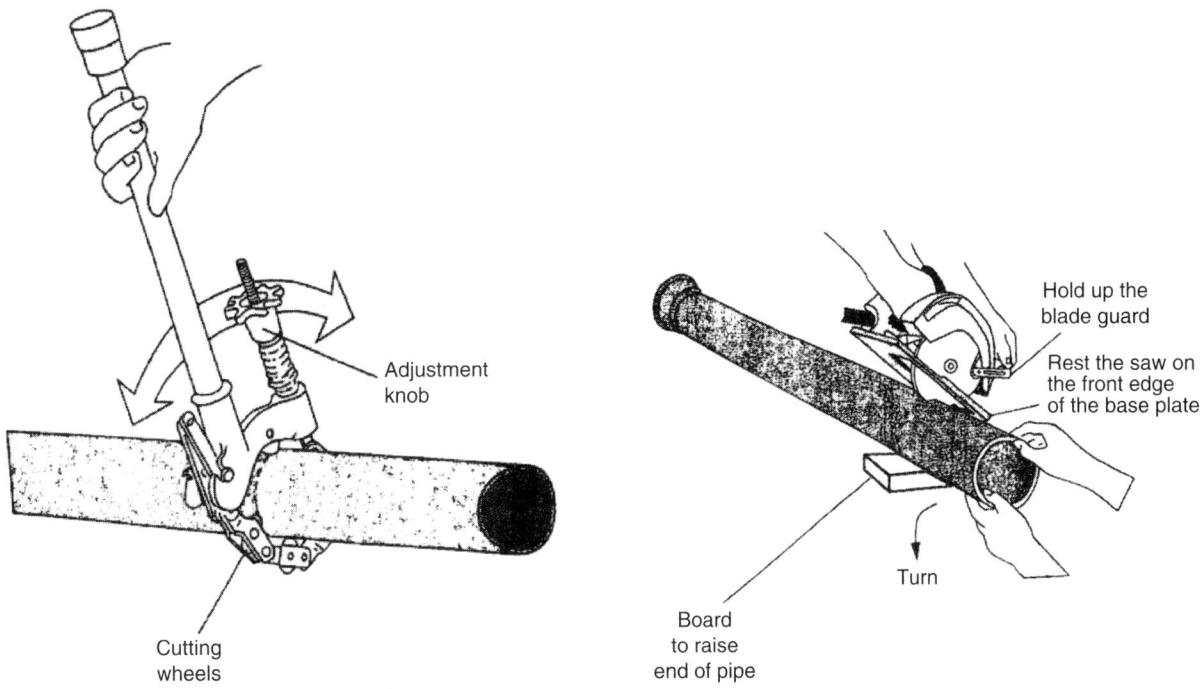

FIGURE 5.1   A chain cutter is wrapped around a cast iron pipe to cut it to length.

FIGURE 5.2   Cutting a cast iron pipe using a circular saw with a metal cutting blade.

lead is melted and poured into the gap between the flange and the side of the pipe. After the lead cools and hardens, it is expanded with a packing iron (sometimes called a tamping iron) so that the lead expands and forms an airtight seal against the two pieces of pipe.

Doing horizontal joints is a bit trickier because the molten metal cannot be poured directly into the flange. That means a joint runner which is shown in Figure 5.5 must be used. The joint runner allows the lead to be directed into the joint. Let the lead set for a short time then it should be expanded as mentioned previously. Most codes now allow the use of flexible unions on horizontal and sloped pipe.

A variety of neoprene or rubber fittings are now made for use with cast iron pipe and they can usually be obtained locally. These include elbows (ells), wyes, and tees. Check with the vendor and permit office first. If used, cut off the flanges and hubs so that each pipe is just a straight cylinder. In some areas you may be able to buy cast iron pipe without the flanges and hubs. Each has a flanged opening just like a plastic or copper fitting. Seal the fitting with a clamp around each flange. Be very careful to support the pipe because the soft fitting will deform and leak if any weight is allowed to rest upon it.

Lightweight, hub-less cast iron pipe is used in some jobs and is a much newer type of cast iron. The connections used for this type of pipe are a type of rubber coupling. Special couplings are designed for joining hub-less cast iron. They have a rubber end that slides over the pipe and fitting. Then a stainless steel band slides over the rubber coupling. A clamp on either end is tightened which compresses the rubber coupling and makes the joint leak-proof. Heavy rubber couplings work well with this pipe.

Rubber Couplings

Heavy rubber couplings can be used with hub-less cast iron. There are special rubber couplings designed to work with hub-less pipe. They can be used to match up cast iron pipe to plastic pipe. However, an adapter is needed for the plastic pipe. The plastic adapter will have one end formed to the proper size to accommodate

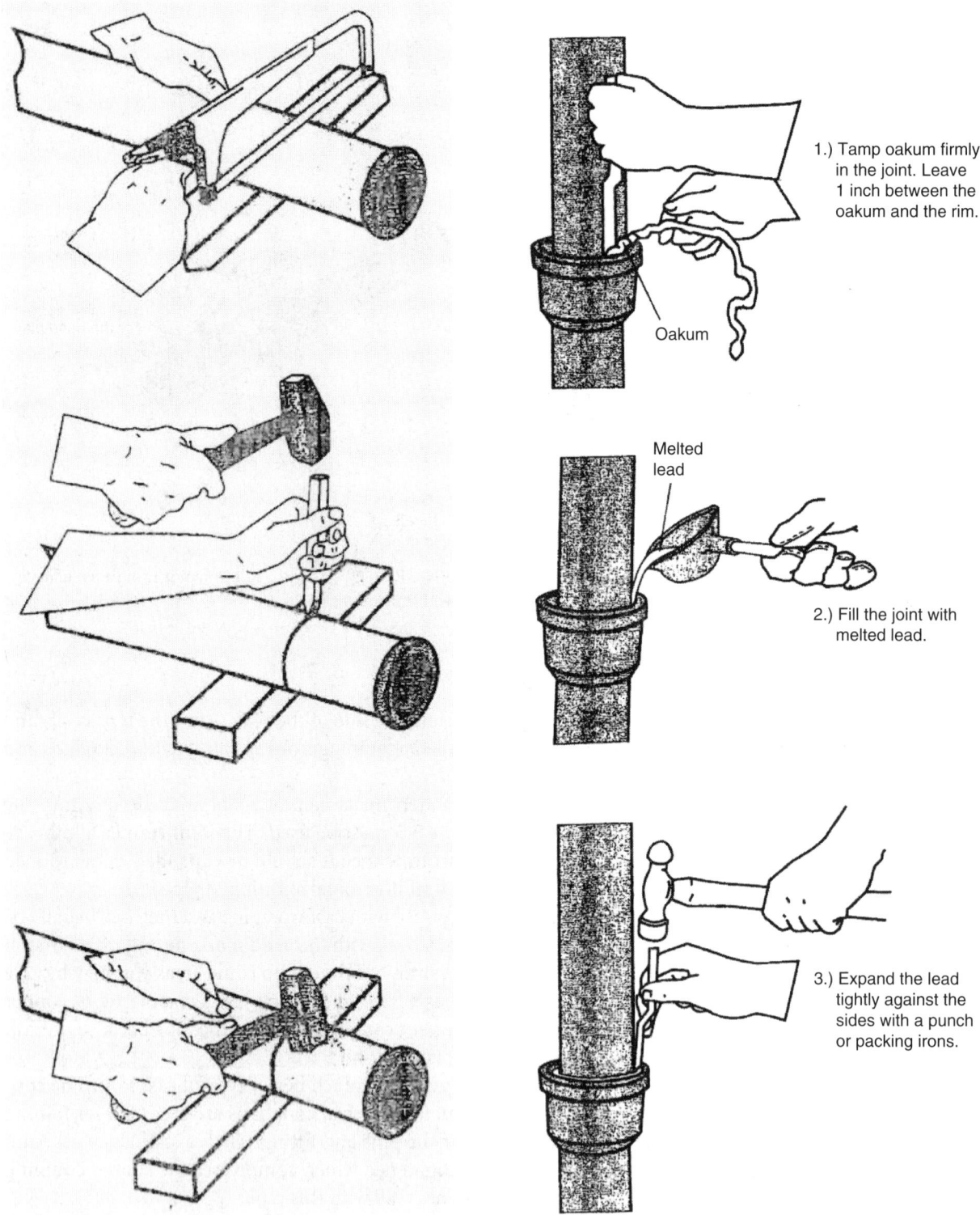

FIGURE 5.3 Cutting a cast iron pipe with a hacksaw and hammer.

FIGURE 5.4 Image showing how to connect cast iron pipe soil stacks.

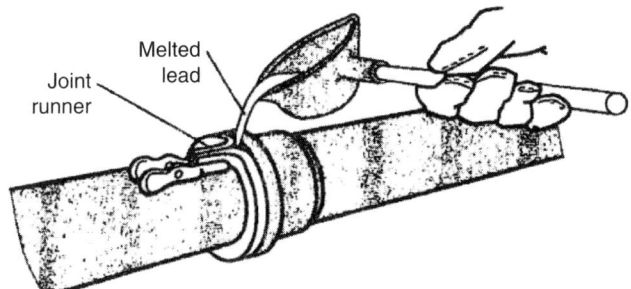

FIGURE 5.5 Connecting cast iron soil stacks horizontally with a joint runner.

the special coupling. Then the adapter is glued onto the plastic pipe and the special coupling slides over the factory-formed end of the adapter.

Threading Pipe

Clean, smooth pipe threads are essential to a good joint and depend largely upon the rake (lip angle) and lead of the chasers, the clearance chip space, and the number of chasers in the die head. The lip angle should vary from 15 to 25 degrees, depending upon the style and condition of the chasers. They should be large enough to allow room for accumulation of chips and at the same time provide a means for lubricating the chasers. Insufficient chip space will cause the chips to clog and tear the threads. The lead of the chaser is the angle which is machined or ground on the leading or front side, to enable the die to start cutting on the pipe, and also to distribute the work of cutting over a number of threads. To secure a good thread, the lead should cover the first three threads. As the heaviest cutting is done by this beveled part, it should have a slightly greater clearance angle than the rest of the threads on the chaser. When regrinding chasers which have become dull on the end, care should be taken to give each chaser the same length of lead or otherwise the work will be unevenly distributed.

The number of chasers with which a die should be equipped depends upon the size of the die. The number recommended for different sizes is as follows:

Size of Die	Number of Chasers
Up to 1¼ inch	4
1¼ to 4 inches	6
4 to 7 inches	8
7 to 10 inches	10
10 to 12 inches	12
12 to 14 inches	14
14 to 18 inches	16
18 to 20 inches	18

Pipe threading dies should be lubricated with a good quality of lard oil or crude cottonseed oil and used in liberal amounts.

WORKING WITH CLAY AND FIBER PIPES

Both clay and fiber pipes are only used for drain or sewer lines outside the home. Most codes require that they should be used no closer to the home than 5 feet. These lines must also be sloped properly to allow waste water to flow into the main sewer. If too much slope is put on the line, the water will run down too quickly and leave the solid matter stuck in the pipe, where it may harden and clog the drain. If not enough slope is put on the pipe neither the water nor the solids will drain effectively. This would result in a slow drain and the possibility that the fluids will back up into the house.

Most codes will require either a ⅛- or ¼-inch of slope per foot. See Figure 5.6. The most common is ⅛ inch per foot. When a greater angle of slope is needed, most codes now require that you use PVC or steel pipe. The standard inside diameter for all these drains is 4 inches.

Vitrified clay pipe is a ceramic product. The clay has been hardened by partial fusion of the clay particles by firing in a kiln. As a ceramic product it is hard and brittle, but extremely durable and resistant to rot or corrosion. It is usually purchased in 2-foot lengths that have a wide flange on one end and a straight cylinder on the other.

Vitrified clay pipe is very difficult to cut. Probably the most accurate method is to use a circular power saw (see Figure 5.2), except that you must use a blade for cutting masonry rather than metal. You can also use a variety of tools to score (cut a shallow line) the mark where the pipe is to be cut. These tools could include a file, a hacksaw, a cold chisel, and even a hatchet. Necessity and lack of planning have often produced unusual solutions. However, you always want to cut the extra length off the straight cylindrical end. Once the line is scored, you can tap around the line with a hammer to break the piece off the main pipe. It will usually break up in pieces rather than as a whole piece.

Once you have all the pieces formed, you must lay the pipe in the trench at the proper slope and then cement the joints. Ordinary masonry cement will work. It is often very tedious and time-consuming to get all the short lengths of pipe sloped properly.

Bituminous fiber pipe is a pipe made from a rot-resistant fiber, such as fiberglass, impregnated with bitumen, a form of tar. It is a lot easier to work with than clay pipe. Fiber pipe is very light in weight, but perhaps the best advantage is that it comes in lengths of 8 to 10 feet. The longer lengths make it easier and faster to lay the pipe.

Both ends of the pipe are tapered so that special fittings, as shown in Figure 5.7, may be used to join the pieces. These fittings are sealed with tar-based cement. Fiber pipe can be cut with ordinary woodcutting tools such as a handsaw or a power saw with a regular woodcutting blade.

After cutting fiber pipe to a shorter length, probably the best and easiest way to join them is with a flexible sleeve coupling. The taper must be cut off the next pipe to get a good seal, though.

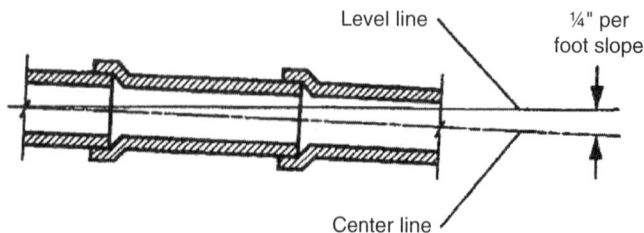

FIGURE 5.6   Sewer drainage line sloping a ¼-inch per foot of length.

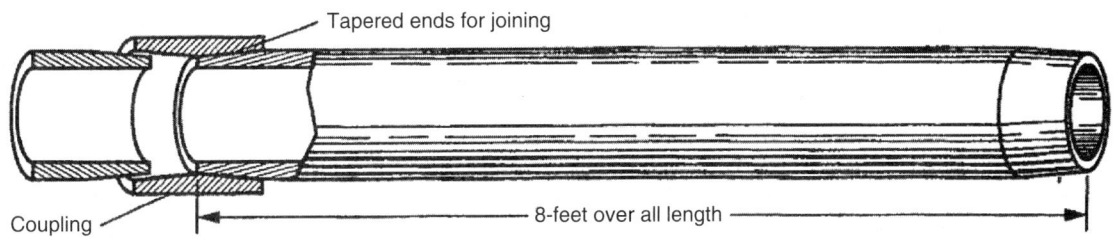

FIGURE 5.7 Bituminous fiber pipe with tapered ends.

WORKING WITH GALVANIZED STEEL PIPE

Repairing a leak or a fitting on galvanized pipe is the same as for any other pipe. However, the major difference is due to the way the pipe is assembled by threading. You cannot unscrew a piece of pipe from the middle. To unscrew the pipe, you must start at one end and disassemble back to the trouble point. Of course, this is simply not possible in most instances. The solution is to cut the pipe.

As a general rule, you need to cut only one pipe. The rest of the problem area can be unscrewed. To fix a pipe leak, follow the procedure shown in Figure 5.8. First, make a cut an inch or so away from the visible leak.

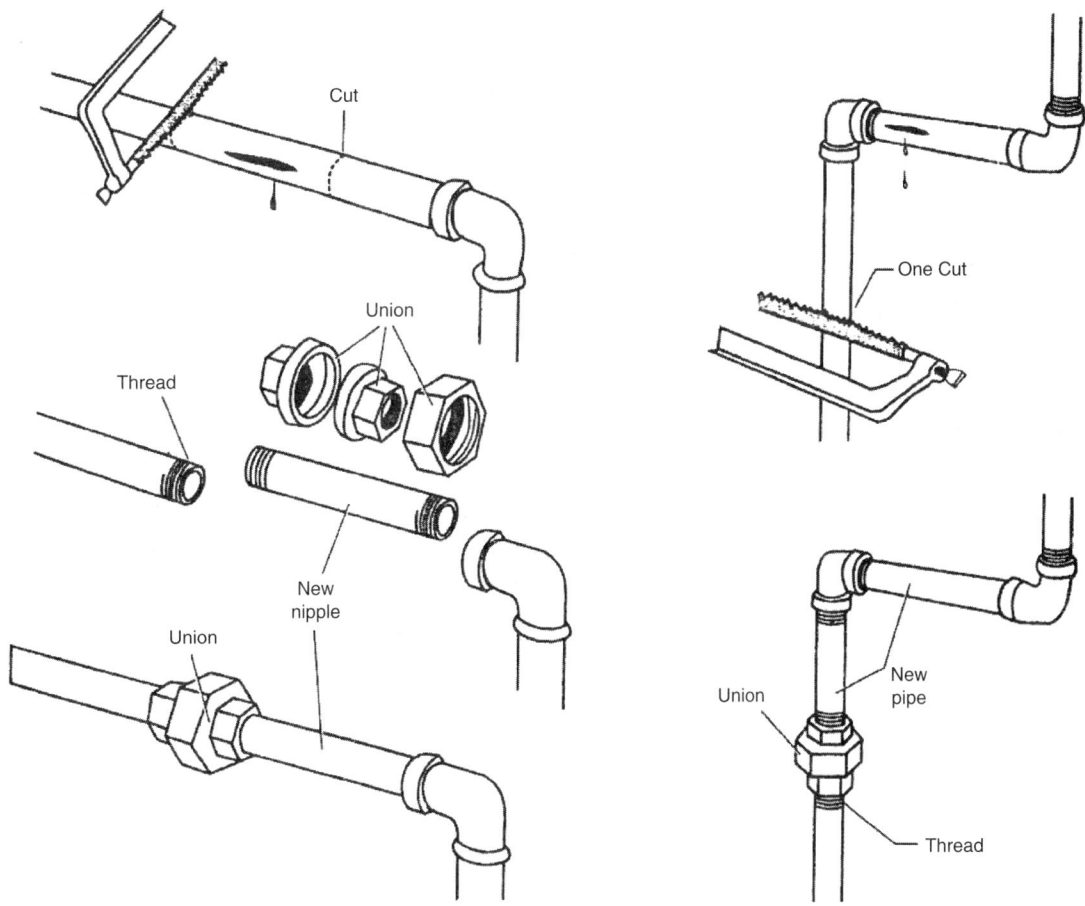

FIGURE 5.8 Repairing a damaged galvanized pipe with a hacksaw and a union.

The handiest tool for this will probably be a reciprocating saw (Sawzall®). A pipe cutter will work if there is room, and/or a hacksaw or mini hacksaw is the next appropriate choice. Then unscrew the damaged part. If it is not very far to a fitting on both sides of the leak, you can replace the damaged pipe with two nipples and a union. If the leak is in the middle of the long pipe, it is best to just cut it out.

A nipple can be used to replace the damaged area as shown in Figure. 5.8 to join the cut sections. However, thread the end of the pipe that was left in place if the length of the pipe to the next fitting is visible, or you can unscrew the pipe and thread it on a workbench or other convenient place. If not, you can use a pipe die and die-wrench to thread it in place. If access is limited, use the ratchet feature of the die-wrench. Then assemble the repair as shown in the illustration.

If you run into a stubborn fitting that just won't unscrew, loosen it first with a penetrating lubricant and tap it gently. If that does not work, try heating it with a torch as shown in Figure 5.9. A minute or so of the heat will cause the fitting to expand away from the pipe. Even just a little expansion will likely break loose the thread enough to unscrew.

To repair a leaky fitting, you usually use an elbow. You must again cut one pipe. The best pipe choice is to pick a short area between the two fittings, as shown in Figure 5.10. In this way, you can use a fitting, two short nipples, and a union for the repair. Cut the shortest length of pipe, as shown. Then pull the pipe out just a bit and unscrew either the elbow side or the straight side. Next, unscrew the other piece. Use the procedure shown in Figure 5.8 to re-assemble the pipe and complete the repair.

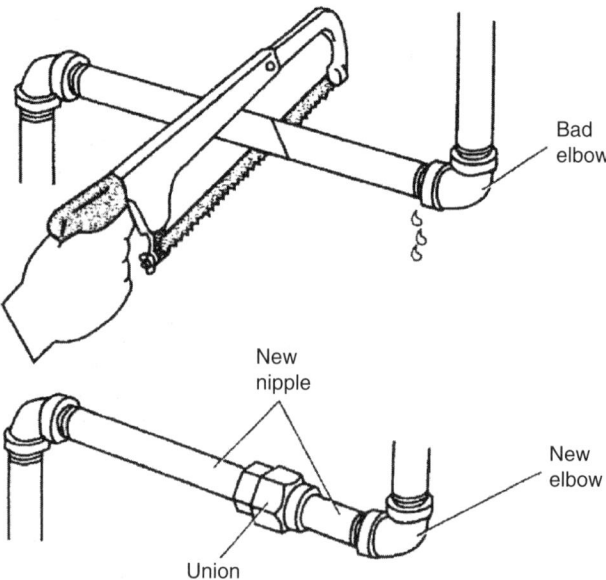

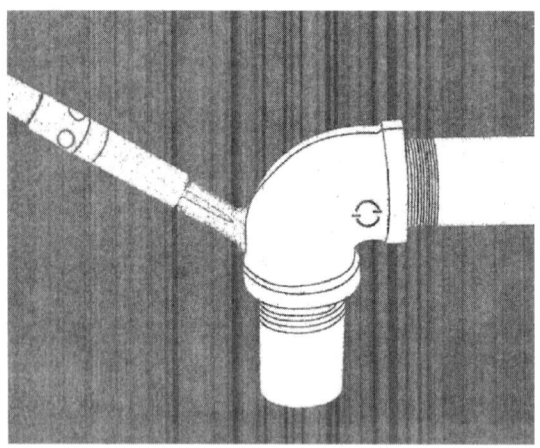

FIGURE 5.9   Heating a pipe fitting for easier removal.

FIGURE 5.10   Installing a union in a broken pipe for ease of future repairs.

REVIEW QUESTIONS

1. When threading a pipe, what is the lip angle of the chasers?

   a. 10
   b. 25
   c. 50
   d. 60

2. What could happen if there is insufficient chip space when cutting threads?

   a. the chasers will rub
   b. threads will cut too deep
   c. threads could get torn
   d. none of these

3. What would be the best way to cut a large pipe made of iron?

   a. table saw
   b. chain cutter
   c. tube cutter
   d. chain saw

4. What is the typical length of a vitrified clay pipe?

   a. 2'
   b. 8'
   c. 12'
   d. 20'

5. What is typically used to fasten vitrified clay pipe together?

   a. rubber grommets
   b. tar
   c. epoxy
   d. masonry cement

6. What is the major component of bituminous fiber pipe?

   a. rope fiber
   b. bituminous
   c. clay
   d. fiberglass

7. What type of adhesive is used to bond bituminous fiber pipe?

   a. asbestos cement
   b. tar-based cement
   c. contact cement
   d. masonry cement

8. What tool is best for quickly cutting galvanized pipe?

   a. pipe cutter
   b. reciprocating saw
   c. chain cutter
   d. hacksaw

9. What may be used to loosen stubborn fitting pipe?

    a. heat   b. ice

    c. liquid soap   d. none of these

10. What should be added to fix a galvanized pipeline so that it can easily be repaired again?

    a. elbow   b. tee

    c. union   d. none of these

ANSWERS TO REVIEW QUESTIONS

1. b	2. c	3. b	4. a	5. d
6. d	7. b	8. b	9. a	10. c

Chapter 6
VALVES

Performance Objectives

After studying this chapter you will (be able to):

1. List the types of valves most commonly used in pipe systems.

2. Discern when to use a relief or safety valve.

3. Discuss how a saddle valve works.

4. Know how a check valve operates.

5. Understand how a globe valve operates.

6. Distinguish the difference between quick shut-off valves.

7. Explain the purpose of a relief valve.

There are a wide variety of valves used to control the flow of liquids and gases and many can be used interchangeably. However, some valves work better than others depending upon the situation. This chapter will cover the main types of valves commonly used in home plumbing as well as in commercial applications. Each type has different construction details and advantages. Figure 6.1 shows a cross section of several types.

CHECK VALVES

Check valves are devices designed to allow a fluid to pass through in one direction only. There are two basic types—the swing check valve and the lift check valve. Working parts of swing check valves consist of a hinged disc or clapper which is free to swing upon a hinge pin in only one direction which should be the direction of the flow of liquid in the pipe line. The pressure exerted by the liquid flowing through the valve lifts the clapper and holds it in an open position. When the flow stops in that direction, the clapper falls back to its original position by gravity, thus preventing back flow. A check valve is shown in Figure 6.2.

Working parts of lift check valves consist of a valve disc so positioned that it is free to rise and fall in a vertical direction. Pressure exerted by the liquid flowing through the valve raises the disc and holds it in an open position. When the flow stops in that direction, the disc drops back to a closed position by gravity, thus preventing forward flow. Any force or tendency to cause back flow only causes the closing action to become stronger in both types of check valves. There are no exterior parts whatsoever on a check valve. Failure of the internal parts cannot be determined by inspection, except by disassembling the valve. Flow through a swing check valve is straightway, as in gate valves. Lift check design, as in globe valves, requires a change in direction of flow through the valve body. A safe rule for choosing check valves is to use swing checks in combination with gate valves; use lift checks with globe and angle valves.

A sump pump will have a check valve so if it turns off, water will not be allowed to flow back in from outdoors and into a basement causing it to flood.

GLOBE VALVES

These valves are the most common type and the least expensive. They use a washer to shut off the water flow, as shown in Figure 6.3. These valves are used as cut off valves for kitchen and bathroom faucets, toilets, and hose bibs (that's the official plumbing name for the faucet outside the house). Globe valves are usually

(a)

Closed

Washer · · · · Partition

Open

Three main types of supply line valves: (a) Globe valve. (b) Gate valve. (c) Ball valve.

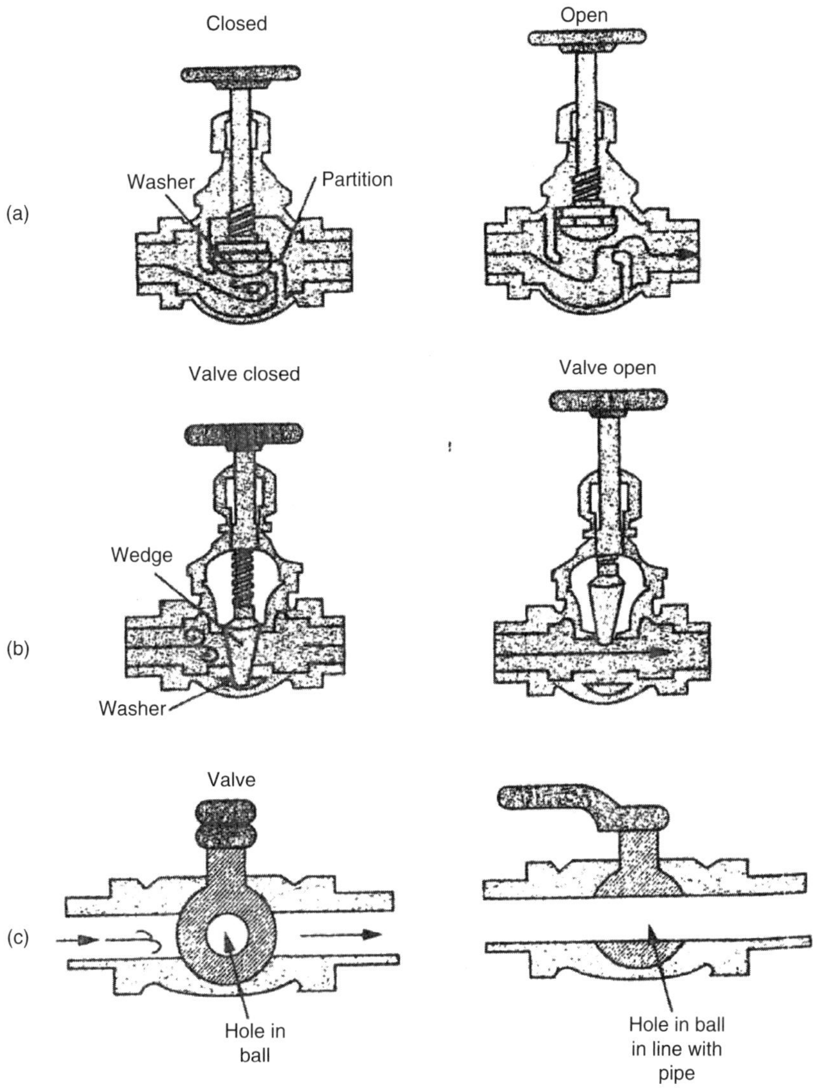

(b)

Valve closed

Wedge

Washer

Valve open

(c)

Valve

Hole in ball

Hole in ball in line with pipe

→ →

FIGURE 6.1 A cross-sectional view of the three main types of valves.

Check Valves

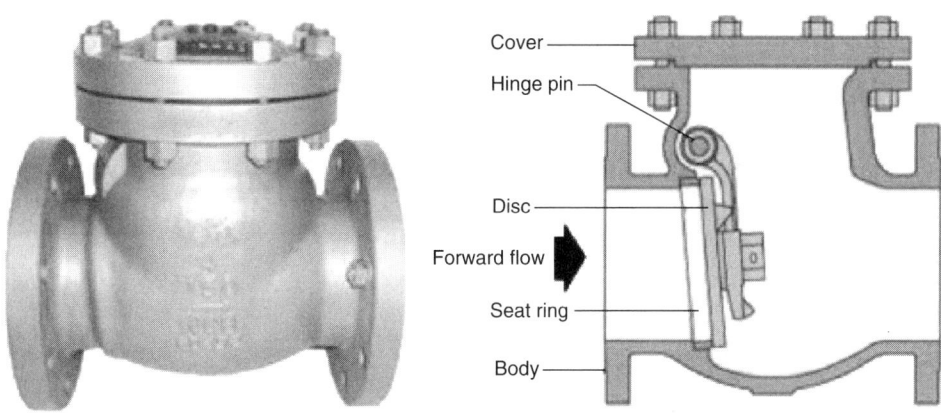

Cover

Hinge pin

Disc

Forward flow

Seat ring

Body

FIGURE 6.2 A heavy duty check valve.

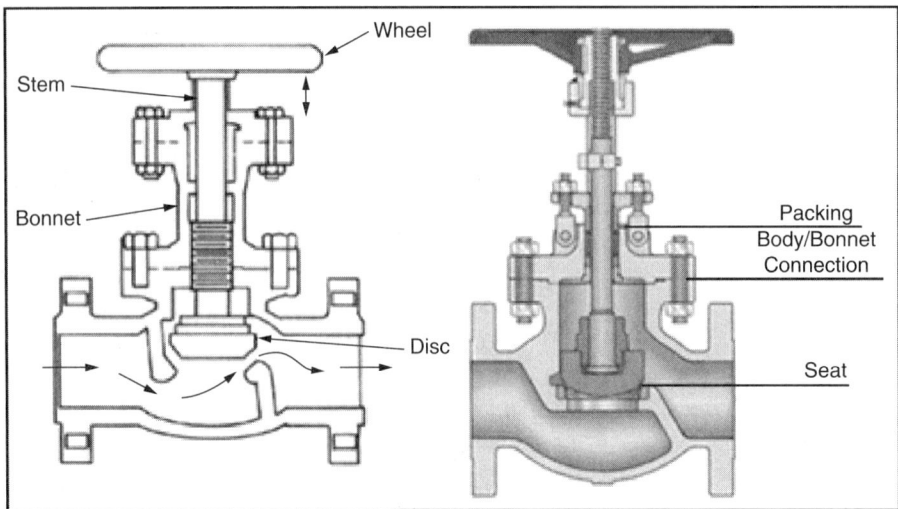

FIGURE 6.3   An example of a globe valve.

made of brass and are often chrome-plated for appearance when used under sinks or any place that is visible. Because they use a soft washer for the seal, they may require maintenance from time to time.

Globe Valves Description and Service Characteristics

Description

The working parts of globe valves consist of a disc which fits over a circular horizontal opening in the valve passageway into which a seat has been fitted or machined; a stem or spindle; a hand wheel; and packing. The same variations that apply to the construction of bonnets, packing glands or nuts, and the type of end connections of gate valves also apply to globe valves as seen in Figure 6.4.

Service Characteristics

Globe valves are well suited to flow regulation (throttling) by hand. Globe valve design causes a change in direction of flow through a valve body, with increased resistance to flow. On liquid lines, pump lines, etc., this may be objectionable, but good for frequency of operation since the valves have short stem travel which saves operator's time. Convenient and quick regrinding features of globe valves make them highly suitable for severe services requiring frequent repair. Installation of a globe valve depends on the type of fluid service to be regulated. If continuous flow is desired in the event that the disc becomes detached from the stem, the valve should be installed in line so that the line pressure is under the disc. If flow stoppage is desired under this circumstance, the valve should be installed with line pressure on top of the disc. A rising stem inside screw type is the simplest and most common stem construction. Stem rises when valve is opened, and thus indicates the position of the disc.

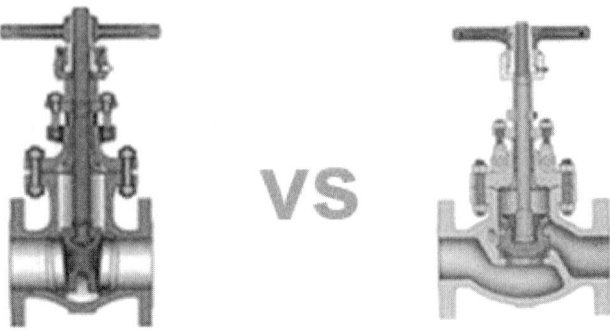

FIGURE 6.4   Gate valve versus a globe valve.

ANGLE VALVES

Angle valves are a type of globe valves and are available in a similar range of constructions. They are used when making a 90-degree turn in a line; an angle valve gives less resistance to flow than the elbow and globe valve which it displaces.

GATE VALVES

This valve features a sliding wedge that cuts across the full diameter of the pipe as illustrated in Figure 6.5. These valves are usually made of bronze. The wedge and its slot are usually machined and mated so that the pressure of the water will tighten and seal the valve in the cutoff position. These valves are larger and longer than check valves and are more expensive.

Gate Valves Description and Service Characteristics

Description

The working parts of gate valves include a solid or split wedge, or gate, which fits into the open passageway of the valve between machined seats; a threaded stem or spindle or hand wheel, and packing. See Figure 6.6.

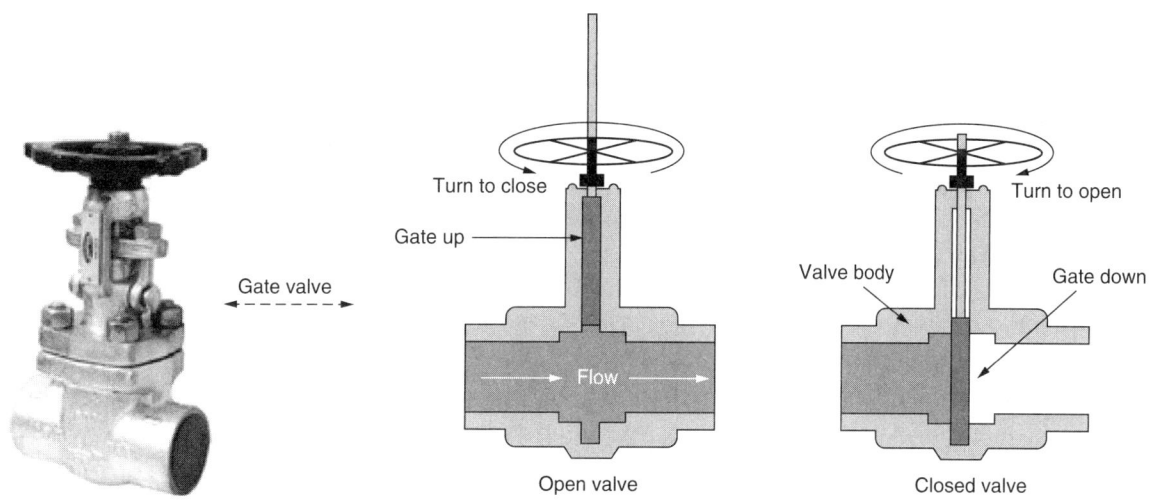

FIGURE 6.5   Parts of a gate valve.

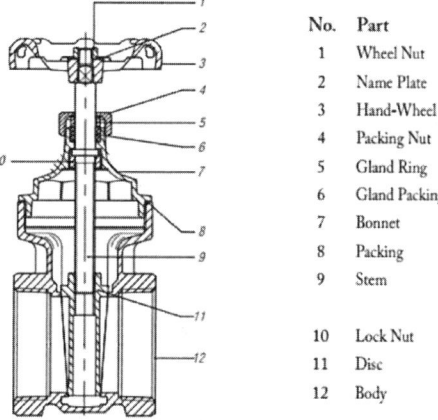

No.	Part
1	Wheel Nut
2	Name Plate
3	Hand-Wheel
4	Packing Nut
5	Gland Ring
6	Gland Packing
7	Bonnet
8	Packing
9	Stem
10	Lock Nut
11	Disc
12	Body

FIGURE 6.6   Diagram of a gate valve.

The bonnet may be of one-piece construction screwed directly to the valve body, or, consist of a union connection screwed to the body. Bonnet may also be bolted to the body, or be constructed as a yoke, exposing stem or spindle. Packing fits around the stem in a recess at the top of the bonnet and is held in place by a packing nut which screws to the bonnet, or by a packing gland which is bolted to the bonnet. A packing gland busing, or follower, installed between the gland or nut and the packing, transmits the force exerted by the packing gland or nut to the packing. Refer to Figure 6.6.

Service Characteristics

Flow through the gate valves is straightway. They are best for lines where unrestricted flow is important—in pump lines, main supply lines, and for stop-valve service. They are suited for service in which valves are infrequently operated, with gate either wide open or fully closed. Gate valves are not considered suitable for throttling services (flow regulation). The split or double disc type should be installed with stem vertical, hand wheel up, as a precaution against possible jamming of the disc spreader mechanism. This type is good for non-condensing gas and liquid services at normal temperatures. The solid wedge type can be installed in any position without danger of the disc jamming. It is recommended as best for steam service and has the highest resistance to pressure strains.

BALL VALVES

The ball valve shown in Figure 6.7 uses the pressure of the water to affect a seal. Ball valves range in size from ¼ to 48 inches. These valves may be made of bronze, brass, PVC, stainless steel, or a combination of these materials. Ball valves are relatively inexpensive and when turned 90 degrees can quickly block the entire flow. They are essentially a ball that fits in the inside diameter of a pipe with a hole in the diameter of the pipe so that when it is open it does not restrict the flow. Ball valves are known for surviving in harsh environments with dirt and other impurities and are best suited for quick shut-off applications. They are not good for fine tuning flow.

SADDLE VALVES

As seen in Figure 6.8, valves are used to connect low-flow auxiliary components to existing water lines without doing extensive adaptation. They are used primarily on rigid copper pipes. The valve is clamped firmly across the pipe, and the needle valve is screwed in to puncture the pipe. The plastic seals around the junction of the valve and the pipe providing a seal to prevent leakage. The most used application is to connect a

FIGURE 6.7   Cutaway view of a ball valve.

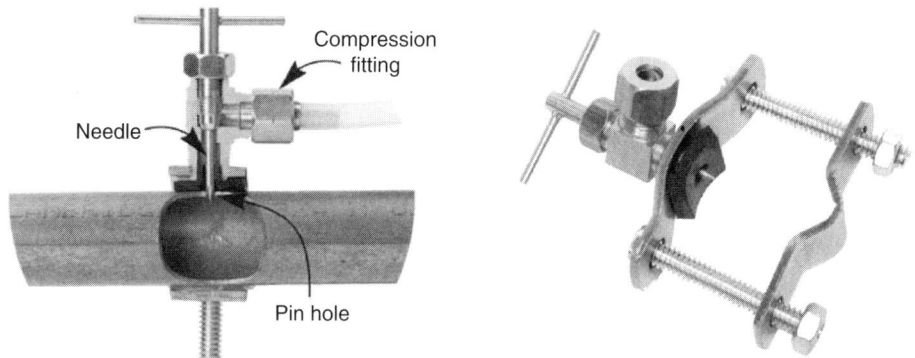

FIGURE 6.8   A saddle valve and how it punctures a copper pipe to divert water to a smaller water line.

flexible copper or plastic line to the icemaker of a refrigerator. It is a quick and easy way to install a small water line from a larger water line without even having to turn off the water.

SILLCOCK

There is one more valve that should be mentioned, it is the sillcock as shown in Figure 6.9. The sillcock is used instead of a hose bib in cold climates where outside access to water is needed, but freezing temperatures will not allow the use of standard valves or hose bibs. Variations of the sillcock are often used as all-weather faucets in yards and farm grounds.

PINCH VALVE

These valves actually act like they are pinching the flow of water by using a flexible rubber tube that is pressed closed when the hand wheel is tightened. They work best for liquids that carry slurries, suspended matter, and solid powders. Refer to Figure 6.10.

FIGURE 6.9   An example of a sillcock valve.

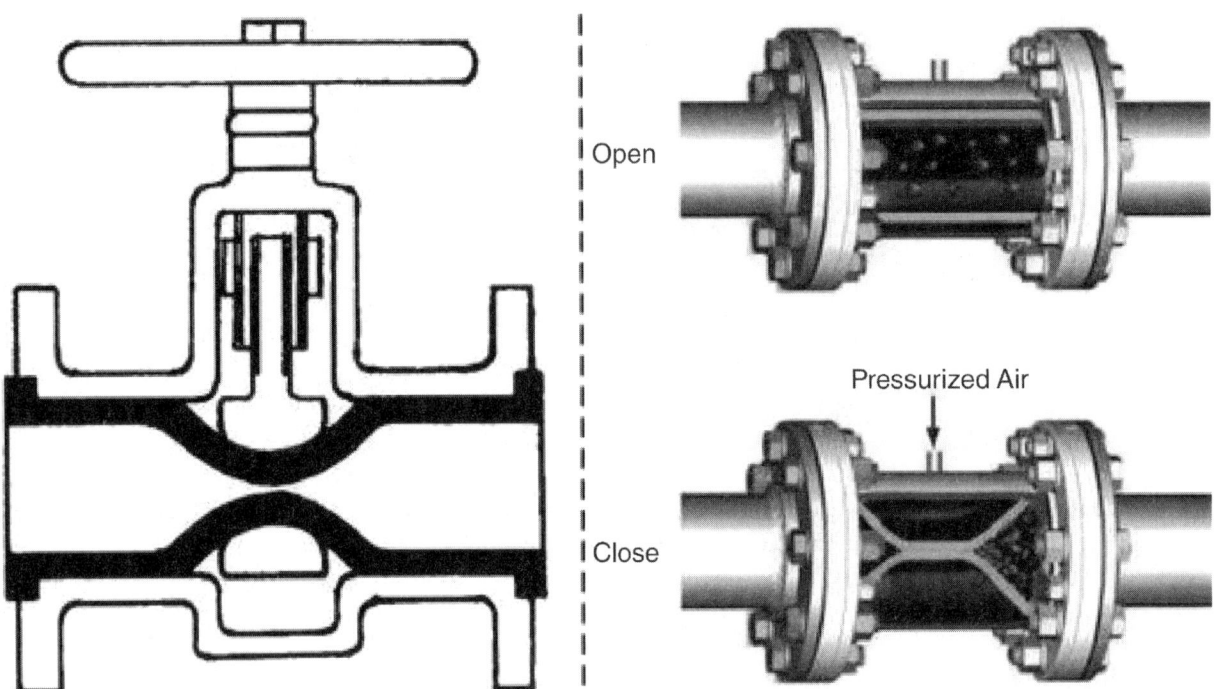

FIGURE 6.10   An example of a hand wheel and pneumatically operated pinch valve.

DIAPHRAGM VALVE

With these valves, a flexible diaphragm seals off the moving parts of the valve from the flow. A diaphragm is a flexible, pressure-sensitive, material that transmits force to open or close a valve. Diaphragm valves are similar to pinch valves, but use an elastomeric diaphragm instead of an elastomeric liner in the valve body, to separate the closing mechanism from the flow stream. Both diaphragm and pinch valves require the elastomeric material to be stretched to open or close the valve; refer to Figure 6.11.

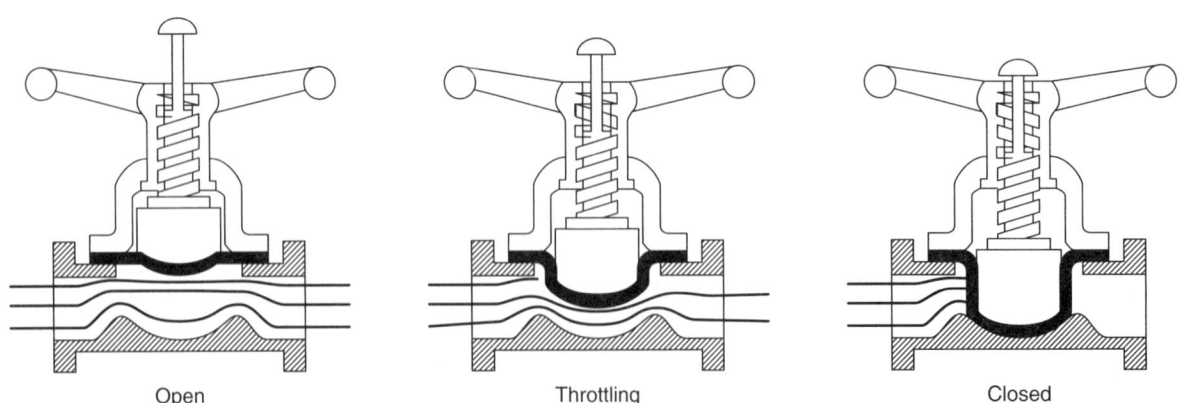

FIGURE 6.11   An example of a diaphragm valve.

FIGURE 6.12 An example of a butterfly valve.

BUTTERFLY VALVES

A butterfly valve is a simple designed valve where a round disc is hinged on a rod where it can be turned 90 degrees to quickly shut off the entire flow. It is a lightweight valve because there are not many parts and good for adjusting flow control. They used to be prone to leaks; however, with newer more durable seating material wrapped around the edge of the disc, as well as the introduction of the more robust double flanged butterfly valve, more of these valves are now used in the chemical, gas and oil, and other processing industries as illustrated in Figure 6.12.

PLUG VALVE

These valves work in the same manner as a ball valve; however, since the center section is either cylindrical or conical shaped with a more rectangular hole in the center, they take up less space and can be used in tight-fitting applications where a typical ball valve could not fit. An example of a plug valve is listed in Figure 6.13.

FIGURE 6.13 A cutaway example of a lined plug valve.

RELIEF AND SAFETY VALVES

A relief valve is spring-loaded, so when too much pressure builds up in a pipe, it will force the valve to compress the spring and release the built up pressure. They are used in applications where noncompressible fluids, such as water or oil, are flowing that do not explode when over pressurized. On the contrary, safety valves are used with gases and steam that could explode when over pressurized as illustrated in Figure 6.14.

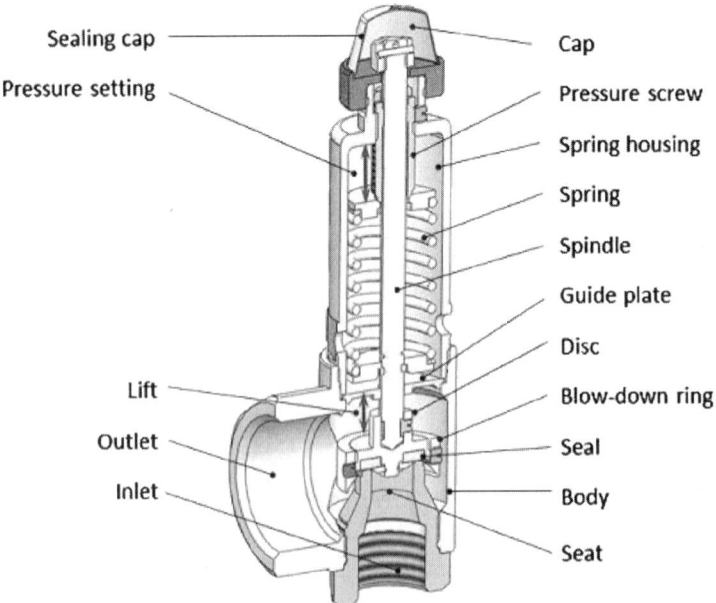

FIGURE 6.14 Cutaway view of a safety valve with list of components.

REVIEW QUESTIONS

1. Which type of valve only lets a fluid pass in one direction?

   a. gate   b. ball

   c. check   d. globe

2. Which type of valve is the most common and least expensive?

   a. gate   b. ball

   c. check   d. globe

3. Which type of valve is typically installed in a sump pump to prevent a basement from flooding?

   a. gate   b. ball

   c. check   d. globe

4. An angle valve is a type of _____ valve.

   a. gate   b. ball

   c. check   d. globe

5. Which type of valve features a sliding wedge to stop the flow?

   a. gate   b. ball

   c. check   d. globe

6. Which type of valve is good for quick shut-off applications with a 90-degree turn?

   a. gate   b. ball

   c. check   d. globe

7. Which type of valve has a center section that is conical or cylindrical in shape and can fit in tight locations?

   a. pinch   b. butterfly

   c. relief   d. plug

   e. saddle

8. Which type of valve is spring-loaded and opens when there is an overabundance of pressure in the line?

   a. pinch   b. butterfly

   c. relief   d. plug

   e. saddle

9.   Which type of valve has a simple design with a round disc hinged on a center rod?

a.   pinch   b.   butterfly

c.   relief   d.   plug

e.   saddle

10. Which type of valve is a rubber tube squeezed to stop the flow?

a.   pinch   b.   butterfly

c.   relief   d.   plug

e.   saddle

ANSWERS TO REVIEW QUESTIONS

1. c	2. d	3. c	4. d	5. a
6. b	7. d	8. c	9. b	10. a

Chapter 7

PIPE MANUFACTURING METHODS

Performance Objectives

After studying this chapter you will (be able to):

1. Know where copper ore is located to make copper pipe and tubing.

2. List the steps involved in making copper.

3. Understand the different purities of copper and the effect on its use.

4. Discuss the advantages and disadvantages of the various methods used to manufacture steel pipe.

5. Identify the labels of various types of manufactured pipe.

COPPER PIPE MANUFACTURING METHODS

To produce hard-working pipes and pipe systems, a number of very different metals are used to their limits. One of these metals is copper. It is used for making tubing of various diameters as well as producing 10-foot-long, very stiff copper pipes. The pipefitter and the plumber are two tradespersons who make good use of the copper in both forms. Copper is a very important engineering metal that has been used by humans for over 6000 years. Used in its pure state, copper is the backbone of the electrical industry, although aluminum is offering competition. Copper is also the major metal in several highly important engineering alloys. Many states still require new housing and offices to be fitted with copper tubing and copper pipe throughout the building.

Most copper ore is in the form of sulfides or oxides, with major U.S. deposits found in Arizona, Utah, New Mexico, Montana, and Michigan. With the depletion of the best domestic ore, deposits in South America (Chile) and Africa have recently assumed great importance.

Even the best ore contains a rather low percentage of copper (0.5%–2%), so the first step is to make a concentration of the actual copper. The concentrated material is floated to produce a product that is about 50% copper. Oxide ores cannot be concentrated by flotation, so an acid leach is used, followed by either precipitation on scrap iron or electrowinning (also known as electroextraction). The concentrate may then be roasted to reduce the sulfur and arsenic. It is then melted with suitable fluxes in a reverberatory or electric furnace, a process known as smelting. Lighter impurities combine and float to the top as a slag. Then, copper, iron, sulfur, and any precious metals that form will produce something known as matte in the lower part of the furnace.

The molten matte (containing about 40% copper) is transferred to a converter that is similar to those used in making Bessemer steel. Air blown through the matte oxides blows out the sulfur and oxidizes the iron that goes into the slag. The product is known as blister copper and is approximately 99% pure.

Blister copper may be upgraded to refined copper by melting it in a furnace and removing the principal impurity, oxygen, in a reducing atmosphere formed by the use of green logs thrust into the melt (poling) or by injection of a reducing gas such as methane. Most blister copper, however, undergoes only a partial furnace refining and is then cast into copper anodes for electrolytic refining. These anodes and thin copper starting sheets, or cathodes, are suspended in tanks containing a solution of copper sulfate and sulfuric acid. An electric current passing through the solution dissolves the copper anodes and deposits refined copper on the cathode. Gold, silver, and other valuables are recovered from the sludge on the bottom of the tanks. The resulting refined copper contains only about 0.07% oxygen and is called electrolytic tough-pitch (ETP) copper. If the intended use of the copper is for the base for an alloy, refinement to low oxygen content may not be necessary and the poling process may be eliminated.

Copper with low oxygen content may also be obtained by two other methods. The first uses an inert-gas atmosphere in the reverberatory furnace and produces an oxygen-free high-conductivity (OFHC) copper. The second procedure is to use deoxidation with a strong reducing agent such as phosphorous or silicon. This approach has the disadvantage that the electrical conductivity is reduced by 10% to 20%. The use of calcium, lithium, or boron as the reducing agent causes less reduction in the conductivity.

Once copper is refined to 99.9% or better, it is melted and the molten metal is poured and cooled (cast) into round long molds call billets. These billets are typically 12 inches or less in diameter and cut at intervals of 2 feet and weigh approximately 400 lbs. The billets are then shipped to a plant that manufactures copper tube. These billets are then reheated to temperatures just over 1500°F to make them pliable enough to be pierced in the center with a mandrel that will form the hollow center of the tube. The pliable copper is then pushed through a smaller die in an extrusion process, just like toothpaste is pushed through the opening of the tube. Because the 12-inch-diameter billet was squeezed into a smaller size of approximately 3 inches the length of it grew to over 80 feet. The tube is then drawn (stretched) through successive smaller dies to reduce the diameter until the desired diameter is achieved, as shown in Figure 7.1.

This drawing process work hardens the copper and makes it stiff or rigid. Typically, if these are sold in straight lengths then the tube is referred to as pipe and just straightened through a few rollers and labeled. If the rigid tube is made into tubing, then it needs to be annealed, or softened, by heating it to 1300°F for a specified amount of time depending on the diameter, gradually cooled, and then coiled and packaged. Tubing or pipe that is used in the medical field or for AC work is thoroughly washed to remove any processing oils and other contaminants and then sealed with removable plastic inserts. All of the manufactured tube is labeled depending upon its specifications as illustrated in Figure 7.2.

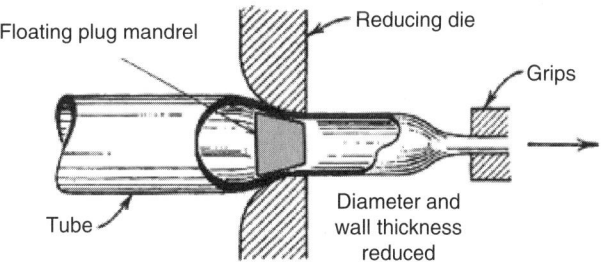

FIGURE 7.1 A diagram of how copper tubing is drawn into small diameter tubing.

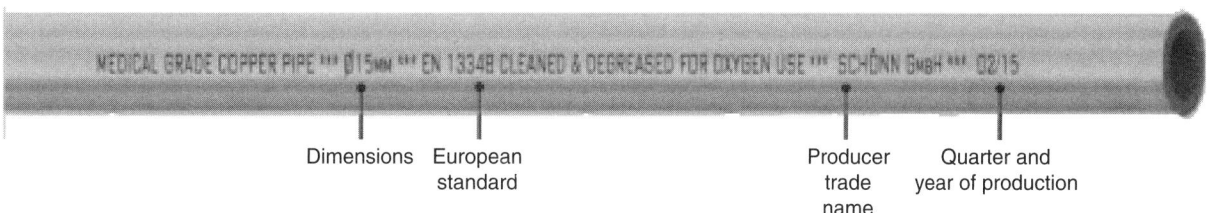

FIGURE 7.2 Illustration of how copper tubing is labeled.

STEEL PIPE MANUFACTURING METHODS

The reason for covering the way in which pipe is manufactured is because some jobs necessitate certain pipe and a pipefitter must be able to distinguish which type of pipe is available by properly reading the label. Furthermore, pipefitters will often have to estimate the cost for a job and should know which type of pipe to use. This is important because only certain types of pipe can meet ASTM standards and codes for high-pressure applications.

Steel is first produced at the steel mill in the form of ingots, billets, blooms, and slabs. An ingot is melted steel poured in a standard tapered, rectangular shape that suits the needs of the customer. It is then further processed by the customer, for example, rolled thinner or remelted into another shape. The ingot is just a standard form that allows it to easily be transported to the customer. It will stack easily on a truck, train, ship, etc. A billet is a length of metal that is round or square in cross-sectional shape that is 6 inches or less in width. A bloom is the same thing only that it is over 6 inches in width. A slab is a long block of metal that is cut to a desired length. It can vary in thickness and width, but it will have a rectangular cross-sectional shape (see Figure 7.3).

Pipe is typically manufactured from billets and slabs depending upon the process. There are four main ways to produce pipe; however, one way has three subclassifications. Three main ways use a slab and the other way uses a billet.

Seamless

The only process that starts with a round billet is the seamless (SMLS) pipe method in which the billet is heated to a temperature where a mandrel with a piercing point can be pushed into the center of the pliable metal to create a hollow tube. It is further finished to the correct size and smoothed to customer specifications. This type of pipe typically comes in sizes from 1/8 to 26 inches in diameter and is available in heavy

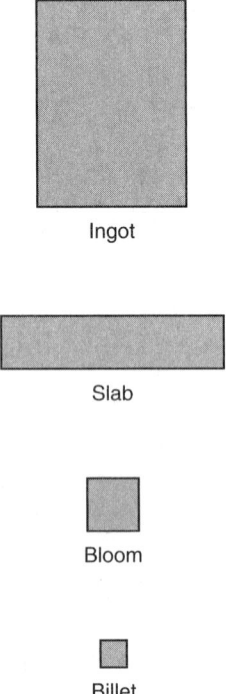

Ingot

Slab

Bloom

Billet

FIGURE 7.3   Shape and size differences between ingots, slabs, blooms, and billets.

wall thicknesses. This process is more expensive than most of the other processes and produces shorter lengths; however, since the pipe is one solid piece with no welds, it can withstand extreme pressures and temperatures and can be used for making any alloy of pipe. This pipe is also referred to as Type S.

Fusion Welded or Continuous Welded

Fusion welded pipe is now better known as continuous welded (CW) pipe since high-density polyethylene (HDPE) pipe is also referred to as fusion welded pipe, and is becoming quite popular, especially in Europe. In this process, a flat metal strip of coiled steel or skelp is heated to approximately 2500°F and then bent with progressive rollers into a circular shape and further pressed together until the hot seam fuses the pipe together. Because the metal is heated, it takes fewer stages of rollers to bend it to the desired shape. In addition, since the metal is hot, it can be pressed and fused together without having to weld the seam. This process is used to make shorter lengths of pipe, especially plumbing pipe that will be threaded. It threads much easier than other pipe because there is no separate welded zone that may be harder than the rest of the pipe. Pipe diameters are smaller and range from 1/8 to 4½ inches. This pipe is also referred to as Type F.

Electric Resistance Welded

The processing of electric resistance welded (ERW) pipe is similar to CW pipe; however, the coiled plate of steel is cold-formed. Because of this, the flat steel has to go through a series of rollers that gradually form it into a cylindrical tube and therefore requires more space to do so. Once the edges of the plate are bent into a circular form, an electric charge is applied to heat the edges so they can then be pressed together, thereby welding it into pipe. ERW pipe is a high speed product that can be made in continuous lengths of over 100 feet. Since the pipe is made with progressive rolls, the pipe has uniform wall thickness. The major disadvantage is that it takes a long time to set up all the progressive rollers so a large number of pipes would need to be produced at a certain diameter to make it cost-effective. However, since it is a high speed process, once the process is set up a large number of pipes can be made quickly with close tolerances at affordable prices. This pipe is also referred to as Type E.

Double Submerged Arc Welded

This is the method that can be further subdivided into three categories for making pipe: (1) U&O, (2) rolled and welded, and (3) spiral welded. All three of these categories use the submerged-arc welding (SAW) process to fuse the pipes together. SAW derives its name from the process wherein the welding arc is submerged in flux while the welding takes place. The pile of flux that is poured on the welded joint protects the steel in the weld area from any impurities in the air when heated to welding temperatures. In fact, it looks like a dimly lit light bulb shining under a pile of dirt so it is possible to look at the welding arc without wearing a welding helmet. The SAW, often referred to as the "subarc" welding process, is an automated process where the welder follows a set of tracks and deposits a lot of material since the arc is buried by a large pile of flux. It is good for running single pass beads that would require several beads if done with another welding process. Because both the inside and outside welds are performed by the subarc welding process, it is referred to as double submerged arc welded (DSAW). The three types of pipe produced by the DSAW process are as follows:

1. **U&O Method.** This process obtains its name from the shape of the dies in the presses used to form the flat strip of metal from which the pipe is made. A U-shaped die first pressed the metal into shape and then an O-shaped die finishes pressing it into a round cylindrical pipe. The seam is then welded using the subarc process from the inside and the outside. This type of pipe is typically cold-formed, so the pipe is very strong and used heavily for gas and oil pipelines. Pipes typically come in 40 feet length. This pipe may be referred to as U&O.

2. **Rolled and Welded Method**. This method is used for short production runs but can be used to make very large diameter thick pipe. There is less lead time with this pipe since it can be made in small batches with a pyramid roll machine that has three rollers, one at the top and two at the bottom. Depending upon the space adjusted between the rollers, the pipe can be rolled into different diameters. Pipe is usually made in smaller lengths of 10 to 20 feet; however, a few manufacturers can make 40-foot lengths upon request. This pipe is also subarc welded on both sides and referred to as R&W.

3. **Spiral Welded Method**. The advantage of spiral welded pipe is that the initial steel plate can be a standard width and turned at a certain angle to make any diameter pipe. The more acute the angle the coil of flat steel is turned, the larger the diameter of the pipe that is produced. It can be made in over 100-foot lengths with 12-foot diameters. Again, it is welded on both sides using the subarc welding process. This pipe is usually referred to as spiral.

REVIEW QUESTIONS

1. Where are the major deposits of copper found in the United States?

   a. Arizona and Montana
   b. Arizona and North Dakota
   c. Alabama and Montana
   d. New Mexico and Texas

2. What is the name of the process that melts out most of the impurities in copper?

   a. electrowinning
   b. Bessemer
   c. Mandrel
   d. smelting

3. What is the term for refined copper that only contains 0.07% oxygen?

   a. copper sludge
   b. electrolytic tough pitch
   c. electrolytic sulfuric pitch
   d. poling

4. What temperature makes copper pliable enough to be extruded?

   a. 840°F
   b. 1300°F
   c. 1500°F
   d. 2500°F

5. At what temperature is copper annealed to make tubing?

   a. 840°F
   b. 1300°F
   c. 1500°F
   d. 2500°F

6. What diameter is common for copper billets?

   a. 2 inches
   b. 4 inches
   c. 12 inches
   d. 36 inches

7. What is the typical weight of a copper billet?

   a. 12 ounces
   b. 12 pounds
   c. 40 pounds
   d. 400 pounds
   e. 4000 pounds

8. What is copper typically refined to that is used in tubing?

   a. 50%
   b. 90%
   c. 99.9%
   d. 100%

9. How long does copper tube become when it is extruded from a 2'-0" billet?

   a. approximately 40'   b. over 80'

   c. over 200'   d. over 500'

10. What is the percentage of copper that is found in ore removed from the ground?

    a. 2%   b. 20%

    c. 40%   d. 99%

11. Which type of pipe is coated with zinc to help resist corrosion?

    a. cast iron   b. black

    c. galvanized   d. lead

    e. brass

12. Which type of pipe is typically coated with lacquer from the factory?

    a. cast iron   b. black

    c. galvanized   d. lead

    e. brass

13. Which type of pipe is primarily used for DWV?

    a. cast iron   b. black

    c. galvanized   d. lead

    e. brass

14. Which type of pipe is chrome plated for appearance?

    a. cast iron   b. black

    c. galvanized   d. lead

    e. brass

15. Which type of pipe is no longer used due to health concerns?

    a. cast iron   b. black

    c. galvanized   d. lead

    e. brass

16. Which type of pipe manufacturing method uses a round billet?

    a. continuous welded   b. seamless

    c. rolled and welded   d. spiral

    e. electric resistive welded

17. Which type of pipe manufacturing method can use standard widths of flat steel to make a range of various diameter pipe?

    a. continuous welded   b. seamless

    c. rolled and welded   d. spiral

    e. electric resistive welded

18. Which type of pipe manufacturing method uses a pyramid-shaped rolling machine?

    a. continuous welded   b. seamless

    c. rolled and welded   d. spiral

    e. electric resistive welded

19. Which type of pipe manufacturing method is cold-formed with progressive rolls?

    a. continuous welded   b. seamless

    c. rolled and welded   d. spiral

    e. electric resistive welded

20. Which type of pipe manufacturing method is well suited for making threaded pipe?

    a. continuous welded   b. seamless

    c. rolled and welded   d. spiral

    e. electric resistive welded

ANSWERS TO REVIEW QUESTIONS

1. a	2. d	3. b	4. c	5. b
6. c	7. d	8. c	9. b	10. a
11. c	12. b	13. a	14. e	15. d
16. b	17. d	18. c	19. e	20. a

—NOTES—

Chapter 8
PIPE FITTINGS AND FLANGES

Performance Objectives

After studying this chapter you will (be able to):

1. Discuss the primary objective of a drainage system.

2. Know what causes a sanitary sewer system to drain or operate properly.

3. List the types of pipe used in drainage systems.

4. Know how fittings are properly used on pipe.

5. Understand how a crosslinked polyethylene (PEX) system works and recognize its advantages.

6. Determine how PEX fittings are attached.

7. Explain how to install a PEX pipe system.

8. Know where and how a CPVC fitting is mounted.

9. Define what PEX-AL-PEX is and how it is used.

10. Understand how drainage systems work.

11. Identify the six main types of flanges.

12. Interpret flange identification markings.

13. Distinguish between the three main flange gaskets.

14. Explain how to properly tighten a flange.

The water supply system terminates at each plumbing fixture in a building of any size. After the water has been drawn and used, it enters the sanitary drainage system. The primary objective of this drainage is to dispose of fluid waste and organic matter as quickly as possible. Refer to Figures 8.1 and 8.2.

Since a sanitary drainage system relies on gravity for its discharge, the pipes are much larger than the water supply lines that are under pressure. Drainage lines are sized according to their location in the system

A building sanitary drain connected to a street sanitary sewer.

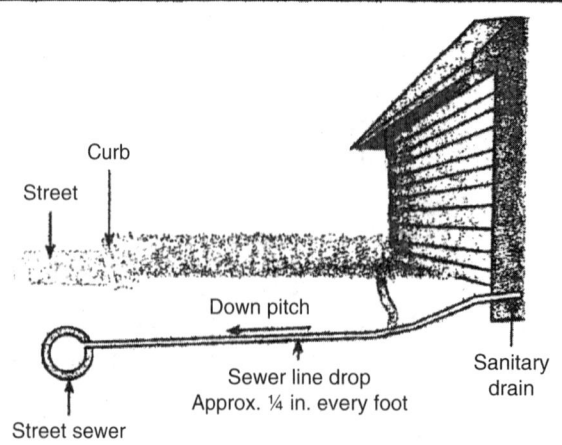

FIGURE 8.1   A sanitary drain connected to a street sanitary sewer (*City of Reno, Nevada, Public Works Maintenance & Operations*).

◆ **ABS DWV PIPE: CELLULAR CORE SCH 40 CUPC ASTM F628   SIZE: 1-1/2", 2",3",4",6"**
   ABS DWV FITTINGS: ASTM D2661 CUPC

◆ **PVC DWV PIPES: ASTM F891 FOAM CORE CUPC   SCH40   SIZE:1-1/2", 2",3",4"**
   PVC DWV FITTINGS: ASTM D2665 UPC

FIGURE 8.2   A sample of ABS and PVC DWV Pipe fittings.

and the total number of types of fixtures served. Always consult the plumbing code for allowable pipe materials, pipe sizing, and restrictions on the length and slope of the horizontal runs and on the types and number of turns allowed. An added drainage line can be added to a main line using a flexible saddle as illustrated in Figure 8.3.

Drainage lines may be of cast iron or plastic. Cast iron, the traditional material used for drainage piping, may have hub-less or bell-and-spigot joints and fittings. See Figure 8.4. The two types of plastic pipe that are suitable for drainage lines are polyvinyl chloride (PVC) and acrylonitrile-butadiene-styrene (ABS). Some building codes also permit the use of advanced wrought iron or steel.

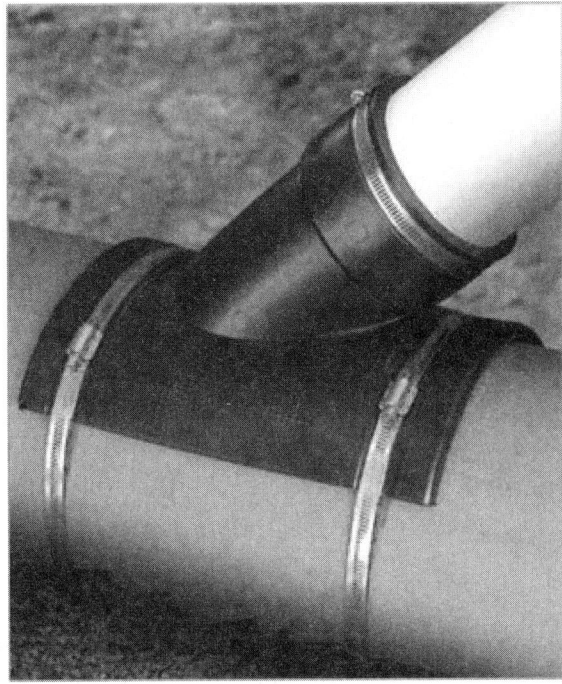

FIGURE 8.3   Flexible saddles (*Fernco, Inc.*).

FIGURE 8.4   Cast iron drainage fittings.

PURPOSE OF FITTINGS

Pipe fittings serve a number of purposes. They couple pipes and join pipes with different diameters. In addition, fittings are used to connect other pieces of pipe to short pieces of threaded pipe with threads at both ends (called a nipple). Elbows, tees, and crosses are used to change the direction of a length of pipe. Caps are used to close open pipes and plugs are used to close open fittings. Bushings reduce the size of openings, while unions make connections convenient and are easily unmade. This includes three-piece screwed unions, where two fittings are screwed to the ends of the pipes being connected and then a thread is screwed onto the first fitting to draw the pipes together by bearing against the shoulder of the second fitting.

FITTINGS

Fittings is the term that generally describes all the various pieces that are used to work the pipe into the desired direction and location. For example, when you need to make a turn in a pipe, or to join two or more pieces together, or to go from one size to another, fittings are generally used. Figure 8.5 shows a number of types of fittings.

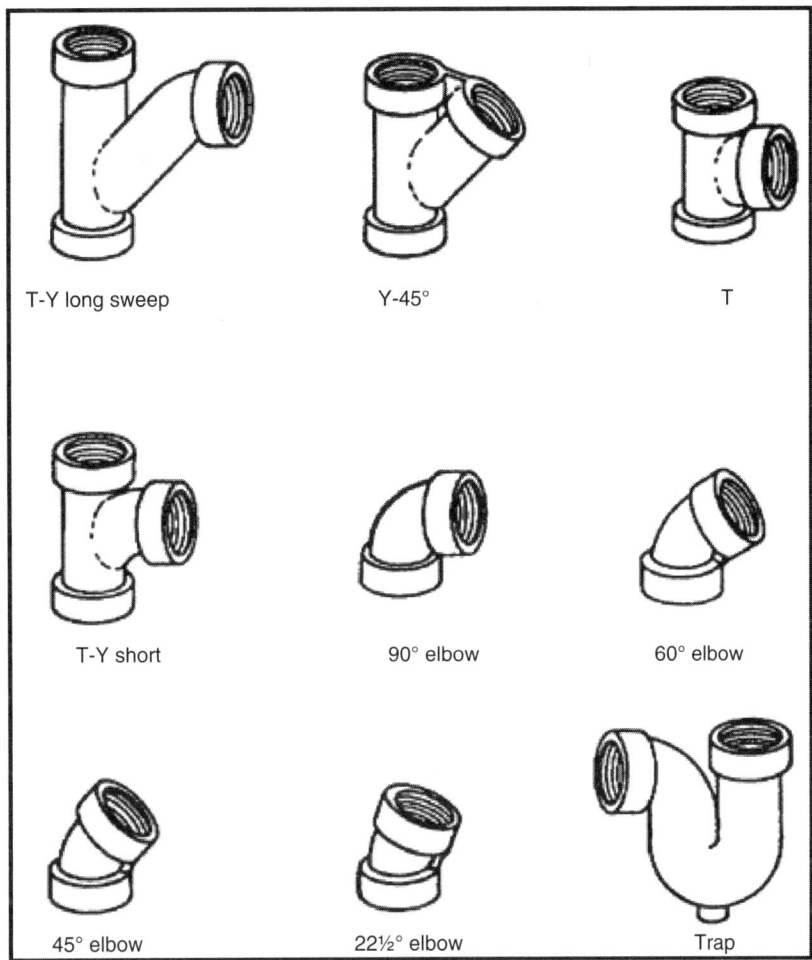

FIGURE 8.5   Fittings allow pipe to change size and direction.

Fittings are used in working with most types of pipe and are available in iron, steel, copper, and plastic depending upon the pipe material composition, and it is almost mandatory that the fittings used match the type of pipe. In other words, if you are working with galvanized pipe, then use galvanized fittings. If not, all kinds of problems may occur—from developing corrosion to not passing inspection. However, oftentimes a threaded fitting is used when connecting a metal pipe with a plastic pipe. Refer to Figure 8.6.

FIGURE 8.6   Fitting to connect metal pipe with plastic pipe. Threaded steel male end and a PVC socket end.

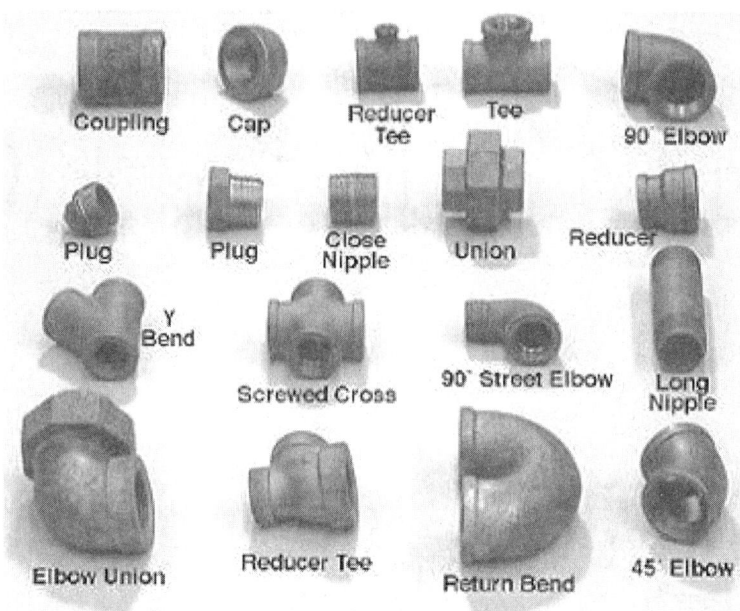

FIGURE 8.7   Basic fitting shapes.

There are a wide range of fittings, as shown in Figure 8.7. Similar shapes are used for steel, plastic, and copper pipes. The range of fittings for cast iron, clay, and fiber pipes is more limited, but 22.5-, 45-, and 90-degree elbows are used as well as 22.5-degree tees and wyes.

  Another point should be made. Fittings for drain, waste, and vent (DWV) are different from regular fittings. As they must carry a large volume of waste and water at a low pressure, they need to provide a more gradual and even change in direction than large regular fittings. DWV fittings are available for every kind of DWV pipe, such as ABS, PVC, cast iron, clay, and fiber. However, the type of fitting should match the type of pipe. ABS fittings should be used only on ABS, cast iron for cast iron, and so forth. The reason is that the cements and sealants for each type are not interchangeable. You cannot join ABS pipe with PVC cement.

The basic group of fittings for galvanized pipe is shown in Figure 8.8. You can work galvanized pipe into almost any location by using just these fittings. However, there is a drawback to using galvanized fittings. A sealant must be used on the threads such as a pipe dope or Teflon tape. The pipe must be assembled sequentially and must be laboriously screwed in place. Heavy pipe wrenches must be used as the bulkiness of the fittings makes their use in tight spaces difficult, and they must be threaded at the factory.

The basic group of fittings for plastic pipe is shown in Figure 8.9 and the basic group of fittings for plastic tubing is shown in Figure 8.10. Rigid copper pipe fittings are shown in Figure 8.11. Both copper and plastic fittings are smaller than steel fittings, making them easier to use in tight places. Plastic fittings are the easiest to use, as they are quickly cemented or welded in place. Copper fittings must be soldered, which involves both heat and time. Soldering fittings also requires some skill.

Fittings for copper tubing (soft copper pipe) are different and are typically compression in nature. Compression unions, shown in Figure 8.12, are commonly available for plastic, galvanized, or copper pipes, and they may be used to join pipes of different materials such as plastic to galvanized pipe.

Any pipe can be moved to any place from any point if the right fittings and nipples are used. Nipples are simply short pieces of pipe, as shown in Figure 8.13.

Threaded nipples for steel pipe can be bought in lengths ranging from 1 inch to 18 inches from most vendors. The nipples are normally sized by the inch up to 6 inches and by 2-inch intervals beyond that length. In some areas, nipples can be obtained that are sized by ½-inch intervals. Fewer lengths are obtainable for

FIGURE 8.8 Basic fittings for galvanized pipe.

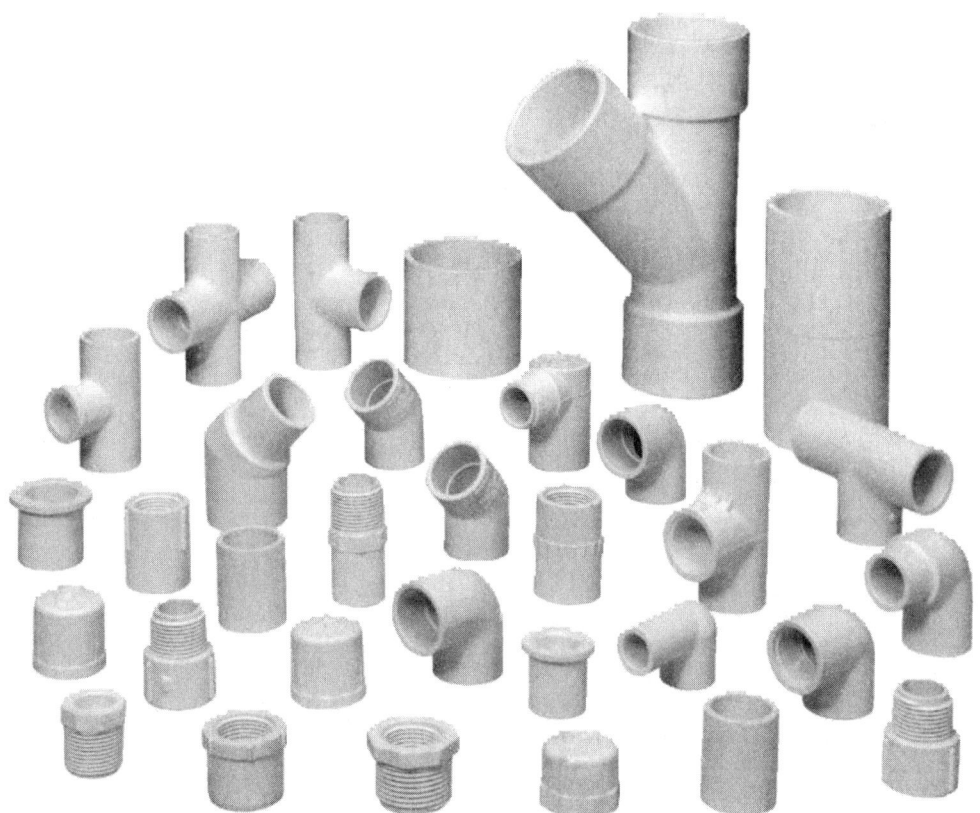

FIGURE 8.9 Basic fittings for plastic pipe.

FIGURE 8.10 Basic barbed fittings used on PE and PB plastic tubing.

FIGURE 8.11 Basic rigid copper pipe fittings.

FIGURE 8.12 A plastic compression union, used to join plastic to metal (NIBCO).

FIGURE 8.13 Various lengths of steel pipe fitting nipples.

plastic pipe, and the sizes are also limited for copper pipe. The reason is that short pieces of plastic or copper pipe are easily cut and cemented or soldered. No unusual tools are required. It is much easier, even for the amateur, to combine nipple lengths than to stop, cut, and thread. Powered-pipe dies can be obtained, but the cost goes far beyond what is reasonable for anything less than the professional use. Many licensed plumbers and pipefitters don't even carry pipe dies with them anymore.

PEX Fittings

PEX fittings are highly flexible and can be easily coiled. PEX resembles polyethylene in appearance. However, because of its crosslinking, it is a thermoset material—it does not melt. Even though PEX tubing will work with all types of PEX fittings, the fittings require specific types of clamps and tools. The barbed end of a PEX fitting is inserted into the PEX tubing and a collar or ring is crimped or pressed onto the end of the tube to make a watertight joint. PEX fittings are approved in all North American plumbing codes for use in hot- and cold-water distribution systems. See Figure 8.14.

FIGURE 8.14   Hot- and cold-water pipe fittings.

Copper Fittings

Copper fittings or copper tubing of various lengths can be utilized more effectively and efficiently by incorporating fittings to attach (couplings), make turns (elbows), or plug up and terminate a pipe or tube (caps). These fittings are all fastened to the pipe with solder and flux. See Figure 8.15.

CPVC Fittings

Fittings for CPVC pipe are made from CPVC and are used in domestic plumbing and fire sprinkler systems. They have a service life of more than 35 years. Outstanding corrosion resistance is another of their features, as well as having a low flame spread and low-smoke emission levels. CPVC fittings are typically tan in color and can handle hotter water temperatures than PVC, so they are used for hot-water lines. Refer to Figure 8.16.

PEX-AL-PEX Composite Fittings

PEX-AL-PEX composite fittings capitalize on the resistance of plastic to chemical corrosion. They can handle pressure as well as any metal can. Because an aluminum layer is laminated between layers of plastic, the tubing is bendable, flexible, noncorroding, and resistant to most acids, alkalis, fats, oils, and salt solutions. Remember that not all PEX tubing has an aluminum layer sandwiched between the plastic. Only the PEX with AL in the name contains aluminum.

FIGURE 8.15   Copper DWV fittings.

FIGURE 8.16   A wide variety of CPVC fitting.

FLANGES

A flange is placed on the end of a pipe so it can be bolted to another standard length of pipe. Flanges allow for easy disassembly of pipe for maintenance, repair, or use for another pipeline. There are a variety of flanges for different applications.

Welding Neck Flange

These are flanges that will be butt-welded to a pipe, fitting, etc., and are easy to recognize because of their long tapered hub. The long tapered hub provides an important reinforcement for use in numerous applications requiring extreme temperatures and pressures. The smooth flange taper is also extremely beneficial under conditions of repeated bending caused by pipeline expansion or other forces as can be seen in Figure 8.17.

Welding neck flanges are bored to match the inside diameter of the mating pipe or fitting so there should be no restriction of product flow. In this way, erosion and any turbulence would be reduced. The tapered thickening hub also provides excellent stress distribution and can easily be x-rayed for flaw detection.

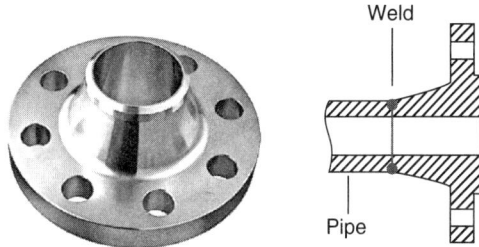

FIGURE 8.17   Welded neck flange.

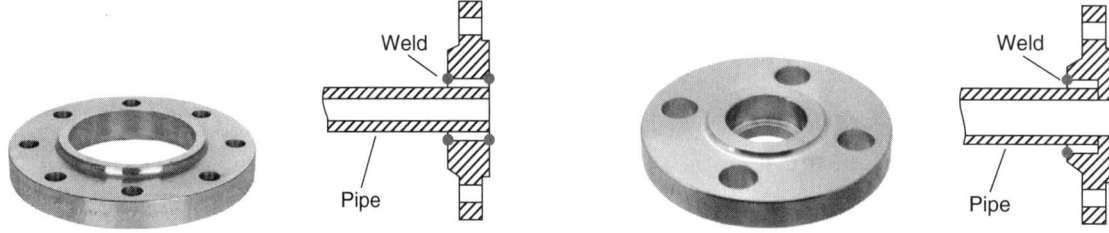

FIGURE 8.18   Slip-on flange.   **FIGURE 8.19**   Socket weld flange.

Slip-on Flange

A slip-on flange is only meant for pipes as it just slips over it and is fillet-welded on the top and bottom. Refer to Figure 8.18. As they do not have a taper hub for extra strength, their life under stress is about one-third of that of a welding neck flange.

Socket Weld Flange

These types of flanges only require one fillet weld on the top once the pipe is placed into it and then given a 1/16-inch gap from the bottom to reduce the residual stress at the root of the weld that can occur during solidification of the weld metal. Socket weld flanges are used for small-size and high-pressure piping that does not transfer highly corrosive fluids. This is due to the fact that these flange types are subject to corrosion in the gap area between the end of the pipe and the shoulder of the socket. A socket weld flange is shown in Figure 8.19.

Threaded Flanges

A threaded flange is joined to pipes by screwing the pipe's male thread onto the female threaded flange as shown in Figure 8.20. The seam between the joint is sometimes welded for extra strength. Threaded flanges are available in sizes up to 4 inches and multiple pressure ratings; however, they are used for smaller size piping in low-pressure and low-temperature applications, such as water and air utility services. Threaded flanges are also a mandatory requirement in explosive areas, such as gas stations and plants, as the execution of welded connections in such environments would be dangerous.

Lap Joint Flanges

These flanges feature a flat face and are always used in conjunction with a stub end. Lap joint flanges look like slip-on flanges except that they have a radius on the back edge to accommodate the stub end. A lap joint flange slips over the pipe and seats on the back of the stub end and the two are kept together by the pressure of the bolts. The use of lap joint flanges in combination with stub ends is a cost-effective solution for stainless steel or nickel alloy pipelines, as the material of the lap joint flange can be of a lower grade (generally carbon steel) than the material of the stub end (which has to match the pipe grade, as in contact with the conveyed fluid). Refer to Figure 8.21.

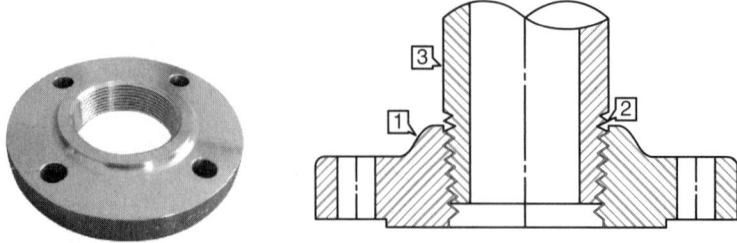

FIGURE 8.20   Threaded flange.

FIGURE 8.21   Lap joint flange.

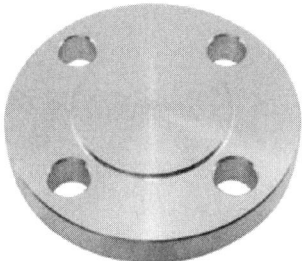

FIGURE 8.22   Blind flange.

Blind Flanges

A blind flange is used to seal a pipeline, fittings, valves, and/or other equipment and block the flow of a fluid since there is no hole in the center. Blind flanges have to withstand extreme pressures. These flanges allow easy access to the pipeline, as they can be easily unbolted as illustrated in Figure 8.22.

FLANGE IDENTIFICATION MARKINGS

Because of the various metal alloys and pressure ratings of flanges, they are typically marked on the edge listing the company that made them, their nominal size, pressure class rating, material designation, melt code, and ring joint groove number if applicable. It should be noted that the pressure ratings depend upon if they are made of cast iron or a steel alloy with each having their specific ASME/ANSI specification. The cast iron flanges are classed from 25 to 800. Steel and steel alloy flanges are available in classes from 150 to 2500.

FLANGE GASKETS

Since pipe flanges need to be bolted together, there is bound to be a leak without some type of gasket. Various types of gaskets are used depending upon the liquid or gas flowing through the pipe. Gaskets must be strong enough to withstand high pressures, various chemicals, and extreme temperatures.

Metallic Ring Gaskets

These types of gaskets are typically made out of a softer grade of metal than the flange it is going to seal is made from. Metallic ring gaskets are best suited for high temperature and pressure applications and are good in corrosive environments. The markings for gaskets typically list the manufacturer, a letter for the type (R = ring gasket), groove number, and a letter for the material type (S = soft iron). A sample of metallic ring gaskets is shown in Figure 8.23.

FIGURE 8.23   A variety of metallic ring gaskets.

FIGURE 8.24   Full face gaskets.

Full Face Gaskets

A full face gasket is designed for use with the flat face of a flange. These gaskets can be made of synthetic rubber, Teflon, and asbestos and metal composition. Synthetic rubber is good for low-temperature and -pressure applications. Teflon and asbestos composition gaskets work well for intermediate temperatures and pressures while metal-asbestos spiral-wound gaskets can handle high-pressure and -temperature applications. Full face gaskets are shown in Figure 8.24.

Flat Ring Gaskets

These type of gaskets are good for raised face flanges and are made of the same materials as full face gaskets. Refer to Figure 8.25.

FLANGE BOLTING

There are two ways to bolt flanges together. One is using a nut with a bolt and the other is using a stud with a nut on either side. The nuts and bolt ratings will depend upon the pressures running through the flange and pipes. The number of bolts, size of bolts, and material that the bolt is manufactured from will be selected according to the size of the flange and how much pressure will be running through it.

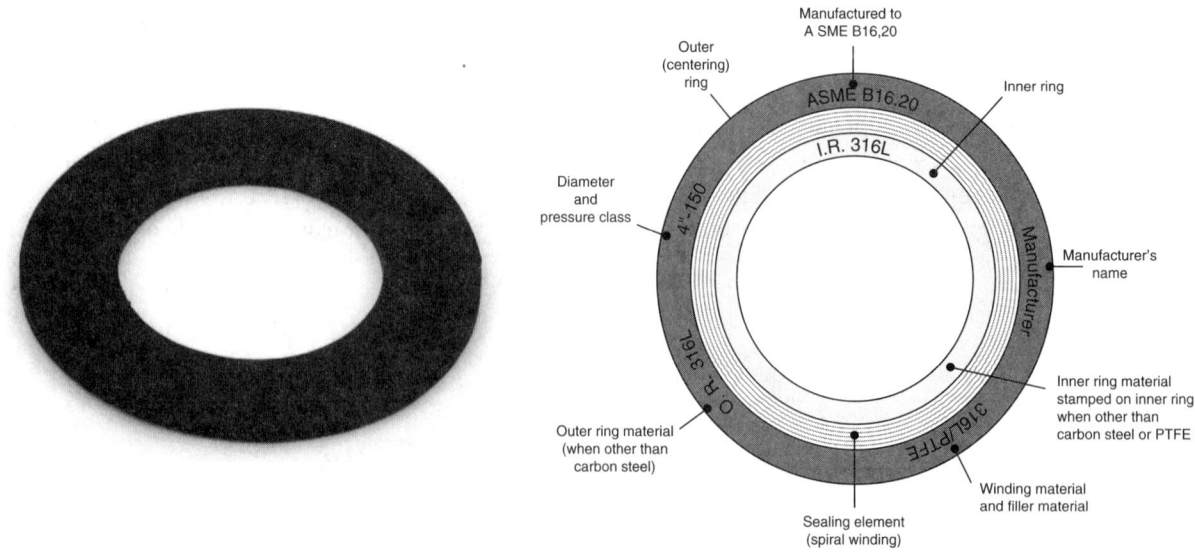

FIGURE 8.25   Flat ring gaskets.

Flange Bolt Tightening Patterns

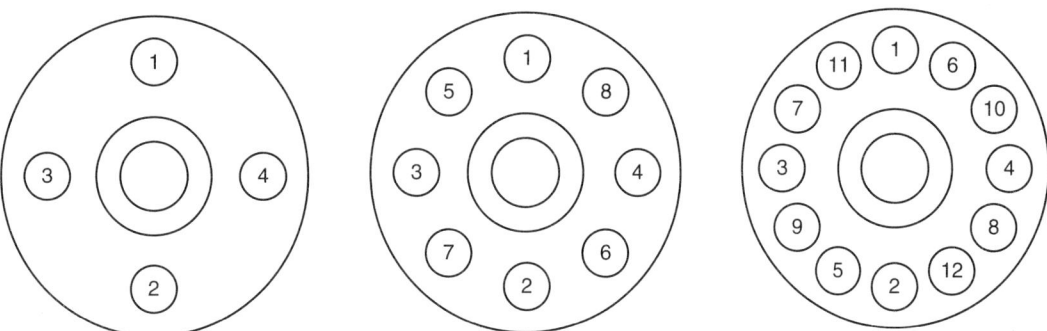

FIGURE 8.26   Tightening pattern for flange bolts.

Before tightening the bolts on the flange, they must first be properly aligned. Thread lubricant should be applied on the bolts to avoid rust and break off during disassembly. Bolts should be tightened snug and then evenly tightened in a criss-cross pattern to the proper torque specification, as illustrated in Figure 8.26. This will ensure that the flange does not get warped and seal properly.

REVIEW QUESTIONS

1. What is the term for various pieces that work pipe into desired directions and locations?
    a. flange   b. valve
    c. tube   d. fitting

2. What is typically used to connect a metal pipe to a plastic pipe?
    a. socket fitting   b. threaded fitting
    c. reducer   d. ell

3. What is the name of a short piece of pipe threaded on both ends?
    a. adaptor   b. coupling
    c. nipple   d. cap

4. A barbed fitting is typically used with _____.
    a. cast iron pipe   b. copper pipe
    c. copper tubing   d. plastic tubing

5. What is the main purpose of an elbow?
    a. make a turn   b. reduce flow
    c. attach dissimilar pipe together   d. strengthen a joint

6. What type of plastic is PEX?
    a. thermoplastic   b. thermoset
    c. thermolinked   d. permobond

7. What is placed at the end of a pipe so it can be bolted to another piece of pipe?
    a. fitting   b. valve
    c. coupling   d. flange

8. Which type of flange would be used when there are conditions of repeated bending caused by pipeline expansion?
    a. slip on   b. welded neck
    c. socket weld   d. threaded
    e. lap joint   f. blind

9. Which type of flange is used to seal a pipeline?

   a. slip on   b. welded neck

   c. socket weld   d. threaded

   e. lap joint   f. blind

10. Which type of flange is used in conjunction with a stub end?

    a. slip-on   b. welded neck

    c. socket weld   d. threaded

    e. lap joint   f. blind

11. Which type of flange must be given a 1/16-inch gap when welding it to a pipe?

    a. slip-on   b. welded neck

    c. socket weld   d. threaded

    e. lap joint   f. blind

12. Which type of flange is used for low-pressure applications and for piping gas stations?

    a. slip-on   b. welded neck

    c. socket weld   d. threaded

    e. lap joint   f. blind

13. Which type of flange is welded on the top and bottom and has 1/3 the life of a welded neck flange?

    a. slip-on   b. coupling

    c. socket weld   d. threaded

    e. lap joint   f. blind

14. What are the pressure ratings of cast iron flanges?

    a. 10–200 psi   b. 25–800 psi

    c. 150–2500 psi   d. 250–10,000 psi

15. What are the pressure ratings of steel and alloy flanges?

    a. 10–200 psi   b. 25–800 psi

    c. 150–2500 psi   d. 250–10,000 psi

16. Which type of flange gaskets are good for raised face flanges?

    a. full face   b. metallic ring

    c. sure ring   d. flat ring

17. Which type of flange gaskets are best suited for high temperatures and pressures?

   a. full face   b. metallic ring

   c. sure ring   d. flat ring

18. Which type of flange gaskets are designed for use with a flat face flange?

   a. full face   b. metallic ring

   c. sure ring   d. flat ring

19. What does the letter S refer to regarding flange gaskets?

   a. silicon   b. smooth

   c. synthetic   d. soft iron

20. What does the letter R refer to regarding flange gaskets?

   a. raised   b. recessed

   c. ring   d. rubber

ANSWERS TO REVIEW QUESTIONS

1. d	2. b	3. c	4. d	5. a
6. b	7. d	8. b	9. f	10. e
11. c	12. d	13. a	14. b	15. c
16. d	17. b	18. a	19. d	20. c

Chapter 9
PIPE SYSTEMS AND REPAIR

Performance Objectives

After studying this chapter you will (be able to):

1. Identify safety equipment located in a factory and/or commercial building.

2. Discuss safety procedures when working with pipe carrying unknown material.

3. Know how to identify and read a color chart.

4. Determine how to make temporary pipe leak repairs.

5. Troubleshoot a piping system and make permanent pipe leak repairs.

PIPING SYSTEM IDENTIFICATION

Identification methods for piping systems have been developed in the past by a large number of industrial plants and organizations. Although these identification methods may make sense to them, the various identification methods have suffered from a lack of uniformity. The lack of uniformity has resulted in considerable confusion, as well as accidents, both for those who change jobs from one plant to another and for those who work at outside agencies, such as municipal fire departments, when called in to assist.

In order to promote greater safety and lessen the chances of error, confusion, or inaction, especially in time of emergency, a uniform code for identification has been developed through the use of color. These standards and legends are shown in Table 9.1.

The classifications for materials transported through a piping system are listed in the following four groups:

1. *Fire protection material and equipment.* Sprinkler systems and other firefighting/protection equipment are included in this group. The identification for this group may also be used to locate such equipment on alarm boxes, extinguishers, fire doors, hose connections, and hydrants.

TABLE 9.1   Color code for labeling industrial pipe.

Industrial pipe marking color codes
Color & application chart

Color	Application
Yellow	Flammable fluids & gases
Red	Fire-quenching fluids
Orange	Toxic corrosive fluids & gasses
Green	All water (portable, boiler, etc.)
Blue	All air (compressed, lab, etc.)
Brown	Combustible fluids & gasses
Purple	Definable by user
Black	Definable by user
White	Definable by user
Gray	Definable by user

2. *Dangerous materials.* This group includes materials that are hazardous to life or property because they are easily ignited, are corrosive at high temperatures and pressures, produce poisonous gases, or are themselves poisonous.

3. *Safe materials.* This group includes materials involving little or no hazard to life or property in their handling. This classification embraces materials that are not poisonous and will not produce fire or explosion at low pressures and temperatures.

4. *Protective materials.* The group's main purpose is to be available through a piping system to prevent or minimize the hazard of the dangerous materials that were previously mentioned.

METHODS OF IDENTIFICATION

To make a positive identification of the contents of a piping system, a lettered legend must be used that gives the name of the material in full or in an abbreviated form. Arrows may be used to indicate the direction of flow. Where it is desirable or necessary to give supplementary information—specifying, say, hazards or the use of the piping systems contents—an additional color must be applied to the entire piping system or colored bands must be used. Refer to Table 9.1 for further details.

There are three main aspects to ANSI/ASME recommendations for pipe content identification: label color, label size, and label placement. These markings allow easy identification for any potential hazards associated with those contents carried in the pipe. The six colors assigned to pipe contents are as follows: yellow, red, orange, green, blue, and brown. Four additional colors (purple, black, white, and gray) can be used as required by a facility; however, these colors must be conveyed to all employees. It should be noted that all labels must be easy to read by anyone entering the facility.

REPAIRING LOW-PRESSURE PIPE DAMAGE

A homeowner occasionally must fix a bad pipe. Damage can be caused by corrosion on either the inside or the outside of the pipe. Pushing or bending a pipe can cause the joints to break or leak. Freezing weather can cause the water in a pipe to freeze and split a pipe or break a joint. On rare occasions, a bubble or fault within the metal or plastic of the pipe can cause the pipe to fail prematurely. Whatever the reason, when pipe failure occurs, water squirts out and repairs are required. See Figure 9.1.

There are different ways of coping with something in the middle of the night, and there are techniques for large leaks and small leaks. First, let's consider the temporary methods that might be used in the middle of the night.

When a pipe fails in the middle of the night, there are several things to consider. There are probably no stores open, so you must do with what you have. The first order of business, of course, is to turn off the water. Then assess the extent of the damage. If you have a drip from a very tiny opening, the easiest thing to do is to wrap the area with several layers of electrical tape. It's also a good idea to add a clamped cover, as shown in Figure 9.2. This can be made from a couple of hose clamps and a piece of sheet metal from a soda can.

Small holes in a metal pipe can be plugged with a sheet metal screw and a soft washer, as shown in Figure 9.3.

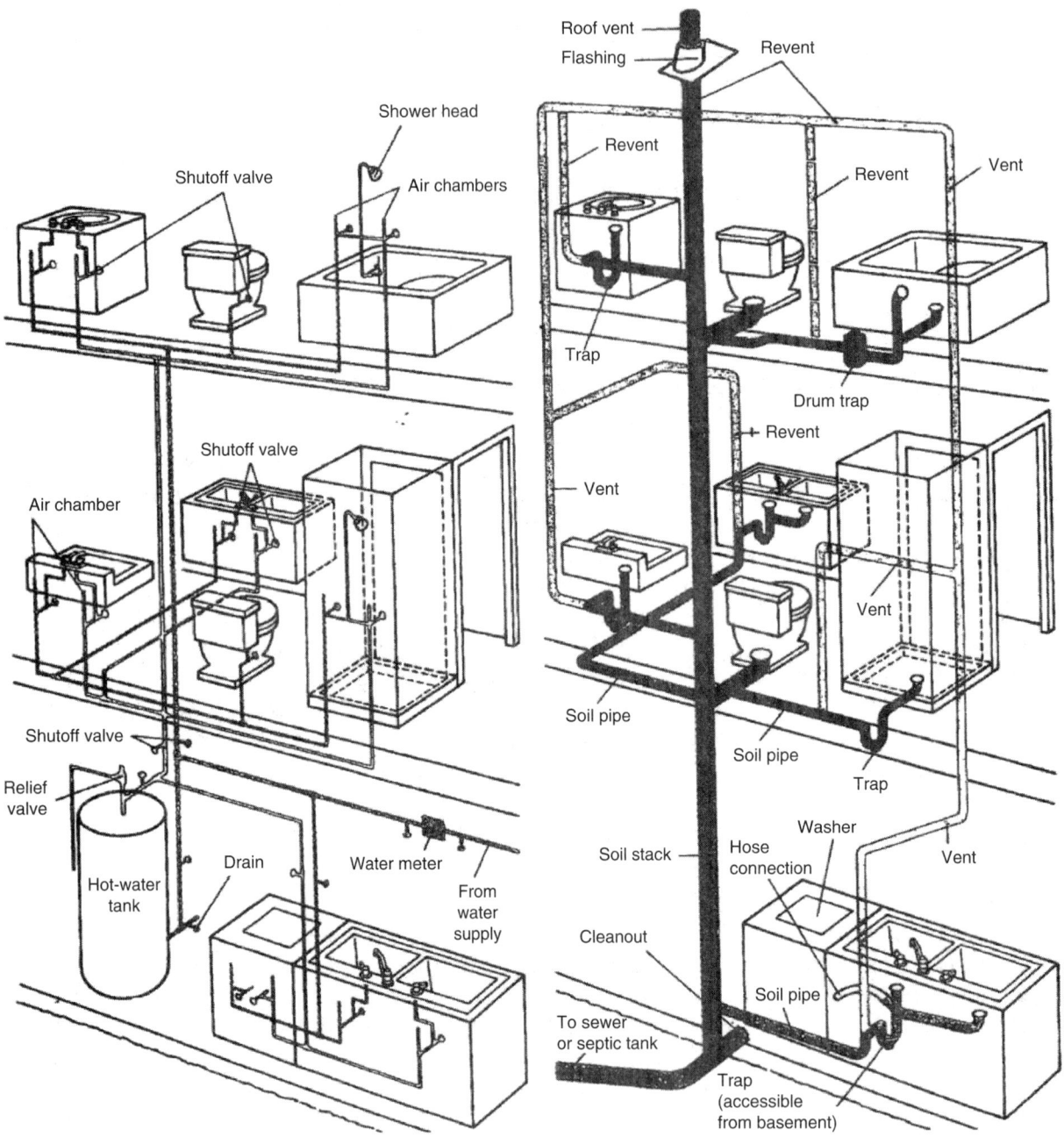

FIGURE 9.1 Typical residential house pipes.

A soft washer can be made from a piece of rubber, gasket material, or a faucet washer. Always use a screw that is short enough so that it will not penetrate the back side of the pipe. Plug the hole with a wooden plug, such as a toothpick or a pencil point, as shown in Figure 9.4. The plug itself may work, but placing a clamp of some kind over the plug works better. There is no need to tighten the clamp very hard; do it just enough to hold the plug in place with the water flowing.

Another trick is to coat the plug with epoxy, wait for an hour, and then turn the water on. Again, it is better to use a clamp over the plug due to the water pressure and the fact that epoxy typically takes 24 hours to fully cure.

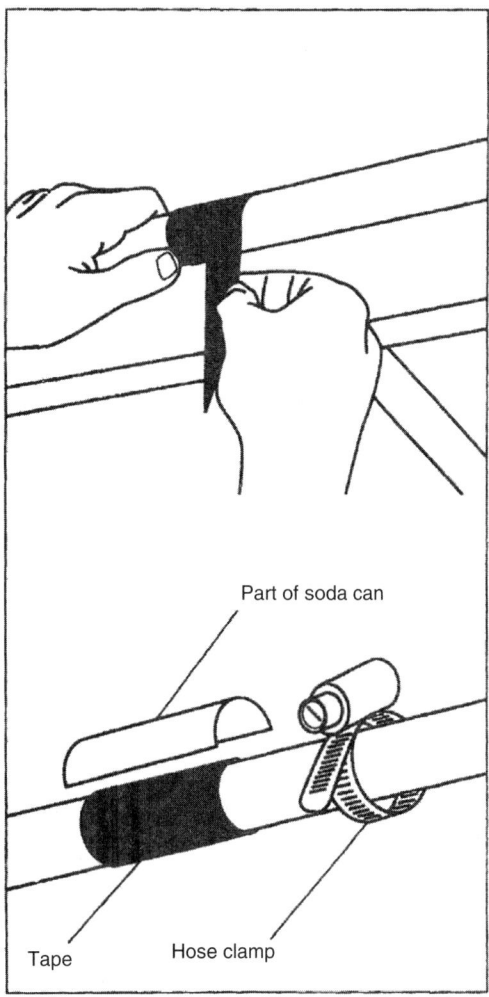

FIGURE 9.2 Wrap electrical tape over a tiny leak. Then clamp metal over it with a hose clamp.

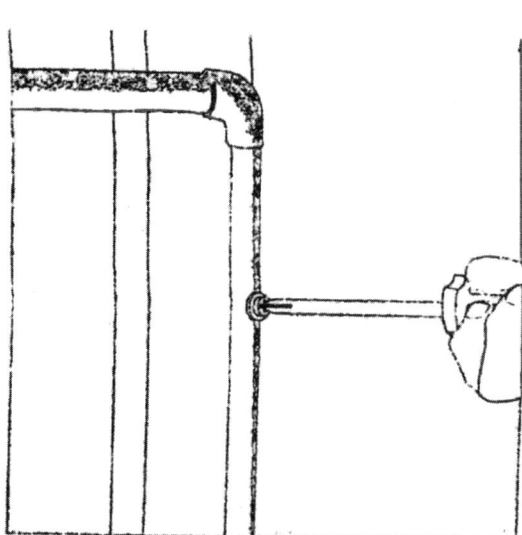

FIGURE 9.3 Plugging a pin hole water leak with a sheet metal screw and washer.

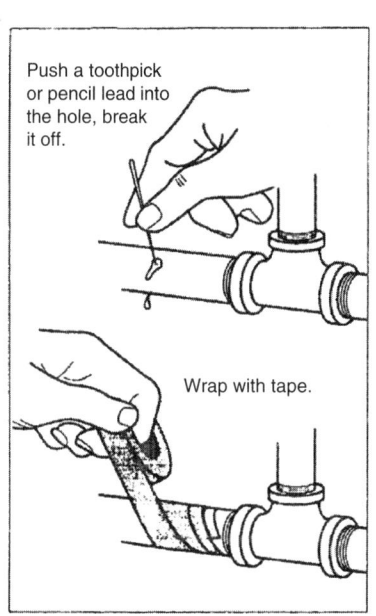

FIGURE 9.4 Plug a small hole with a toothpick. Then wrap it with tape.

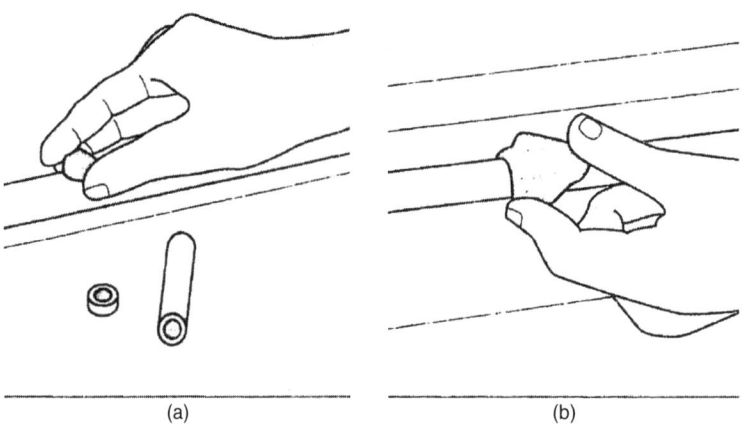

FIGURE 9.5 Plumber's epoxy putty works well. Cut off a small amount, knead it, and press it into the hole.

There are several types of epoxy available; however, the handiest is similar to a small roll of putty; see Figure 9.5. The epoxy resin is in the core of the roll and the hardener is on the surface. First, clean the surface around the leak with an abrasive and make sure it is dry. Then, simply break off a small piece of putty, knead the hardener with the resin, and press it into and over the leak. Allow it to harden before you turn on the water. Follow the directions on the label, because some epoxies harden faster than others. Then check the patched area to be sure it doesn't leak.

Still another type of repair is possible. Actually, there are two versions of this type: a factory-made one and a hastily rigged middle-of-the-night home repair version. Both are shown in Figure 9.6.

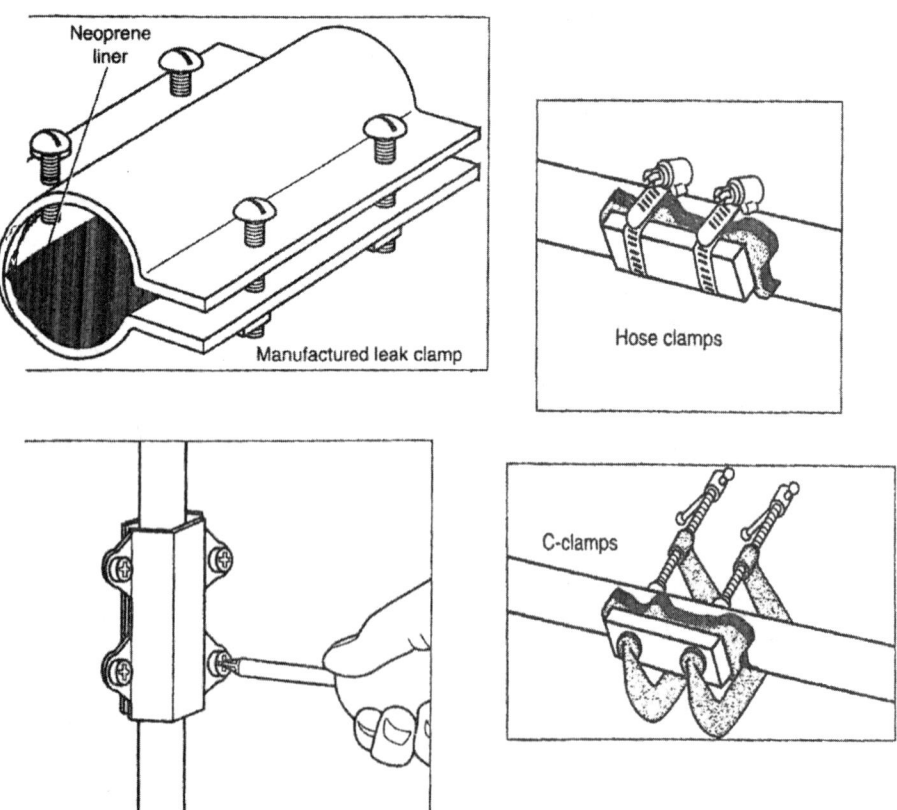

FIGURE 9.6 Factory made and improvised leak repair methods.

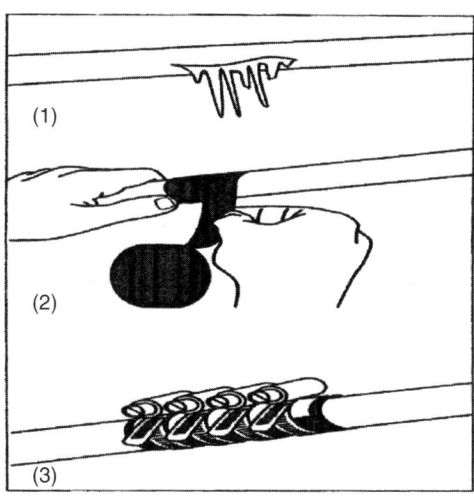

FIGURE 9.7 Temporary patches over large leaks can be done with tape, tin, and hose clamps.

For a lengthy leak, a version of the rubber and the soda can strategy can work. The key to making this work is to use several hose clamps along the length of the patch, as illustrated in Figure 9.7.

What to Have at the Ready

To end this section on temporary fixes, here are some suggestions of what to keep on hand. Epoxy putty will keep for months, if not years, and is highly recommended. Several extra hose clamps are also on the list. Obtain clamps for diameters that are about 1 inch more than the pipe diameter. Remember that the outside diameter of the pipe is greater than the official size of the pipe. Therefore, if the pipe is a 3/8-inch galvanized supply line, which has an outside diameter of about 1 inch, then use clamps good enough for up to 2 inches.

Obtain some rubber or neoprene scraps. If no scraps are available, buy the smallest, cheapest bicycle tire you can. It won't cost much makes excellent patch pieces. Use gasket material if necessary, if an automotive supply house is nearby. Then, add a couple of pipe leak clamps, shown in Figure 9.6. If these are not available, consider a couple of C-clamps of sufficient capacity.

Thawing Pipes

Sometimes pipes freeze in places that can't be seen. It's common for pipes to freeze around basement walls and in the outside walls of a house. An open space can also allow the pipes to freeze. In the initial stages, a pipe will usually freeze in one place first, with the freezing action spreading from there. At this stage you can thaw the frozen area without pipe damage. Pipes carrying substances other than water will, of course, freeze at a different temperature from water, depending upon their freezing point. Freezing pipes are a problem because water expands when it freezes. The freezing action can eventually result in pipe damage, as previously mentioned. Anything that applies heat on the pipe will work to thaw it. One warning, however, don't use a torch or any type of high heat source on plastic pipe. See Figure 9.8. You can pour hot water on the pipe. Wrapping the pipe in rags to hold the water will help retain the heat. Hair dryers and lamps will work. While a heat lamp is good, any electric lamp will work. Heating pads and heating strips are good solutions, too. Even a wrapped pipe with a lit lamp nearby will serve to gradually melt the ice inside the pipe.

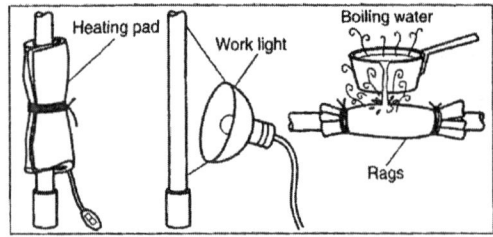

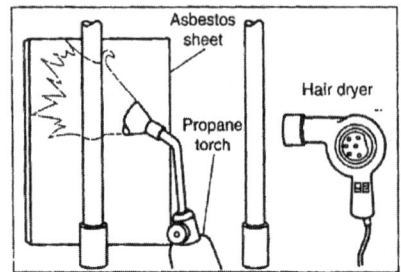

FIGURE 9.8   Ways of thawing a frozen pipe.
Do not use a torch on plastic pipe.

PERMANENT PIPE REPAIRS

Making a permanent repair to a pipe or a system requires a little more attention to detail than making a temporary repair. To make a permanent repair, the faulty section must be removed. There usually are two reasons why a pipe or system will fail: (1) there is a break in a pipe and (2) there is a bad fitting or joint. The solution: To fix a broken pipe, cut out the damaged portion and replace it. In order to fix a bad fitting, cut at least one of the pipes leading to it and remove the faulty fitting. The method will vary depending upon the accessibility of the pipe. Ways to make the repairs are presented below for each type of pipe.

Repairing Copper and Plastic Pipe

Repairs for both copper and plastic pipe are done in almost the same manner. The main difference is that one is soldered and the other is cemented. To fix a pipe leak due to a break in the pipe, cut an inch or so past the leak on each side. A tubing cutter works well where there is room. If not, a reciprocal saw or a mini hacksaw is a good tool to use. Do not try to cut near the damaged area of the pipe. There may be more damage than seen. Instead, cut a little extra out of the good part of the pipe to be sure all the damaged area is removed. Get rid of the burrs and then make sure the distance needed to replace the cutout part is sufficient, as illustrated in Figure 9.9.

There are two options. The first is to use slip couplings, as shown in Figure 9.9. Slip couplings are fittings that do not have a central ridge in them. Thus, they slip well past the break onto the good pipe. Put the new pipe section in place and slide the fittings back over the joint and fix them in place. It is a good idea to mark the place where the fittings should be fixed. Once you put slip couplings over the joints, it is difficult to see exactly where they are in respect to the ends of the pipe. A mark to align each fitting is a big help. Slip couplings are becoming scarce and are simply not available in all regions.

The second method of repair is to use a union, as shown in Figure 9.10.

Remember that there are unions for copper pipe and for plastic pipe. Each type has a shoulder fitting on each end of the union and can be soldered or cemented to the pipe. Again, cut away the damaged pipe and a bit more. Next, measure the total distance that must be replaced, allowing for the width of the union and the shoulder depth of the fittings. Finally, cut nipples of the approximate length and fix them in place. Take apart the union. Slip the threaded nut onto one pipe and attach the unthreaded part of the pipe. Then attach the threaded part of the union to the other side of the pipe, as shown in the figure. Screw the two parts together and complete the repair. Unions allow a pipe to be taken apart easily if needed in the future.

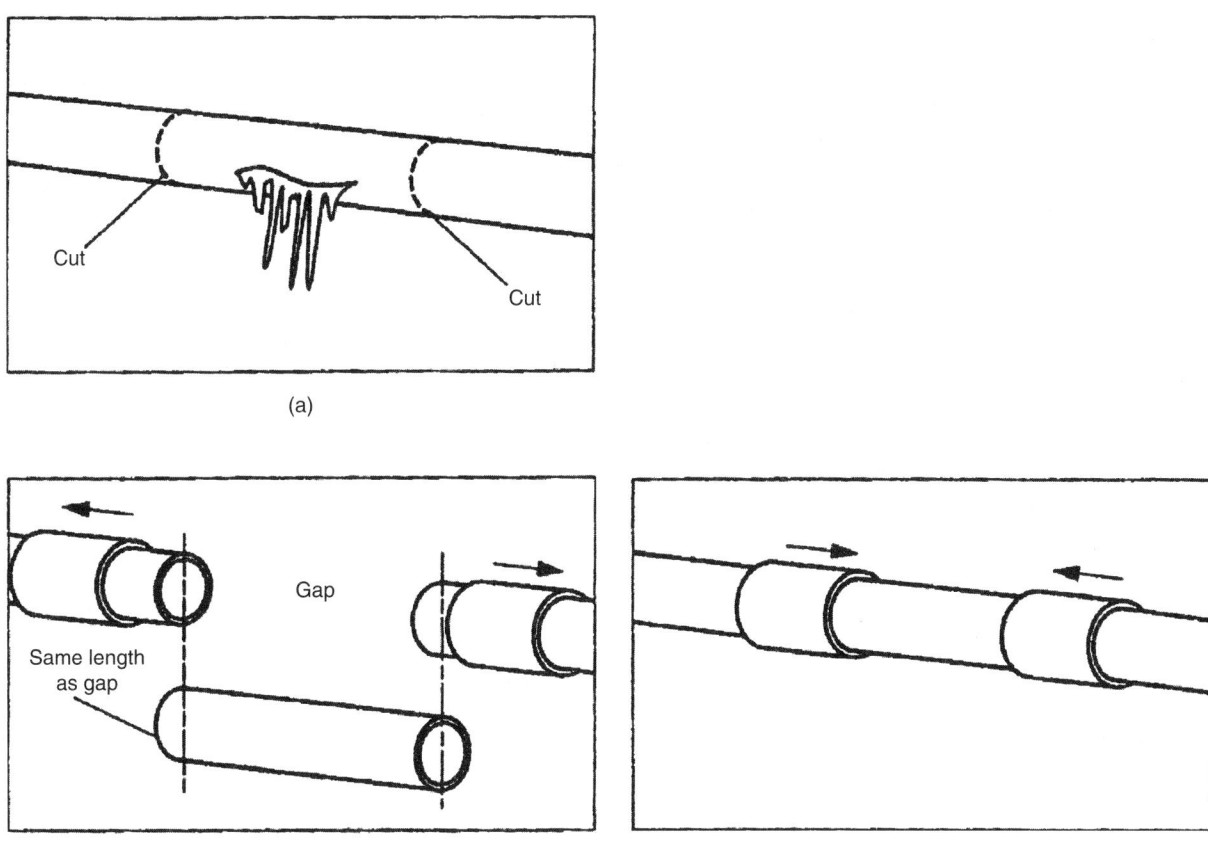

FIGURE 9.9 Fixing a leak with slip fittings: (a) Cut out the section of copper or plastic. (b) Cut a new piece the same length as the gap. Slide slip couplings on each side. (c) Apply cement or solder and slide in place.

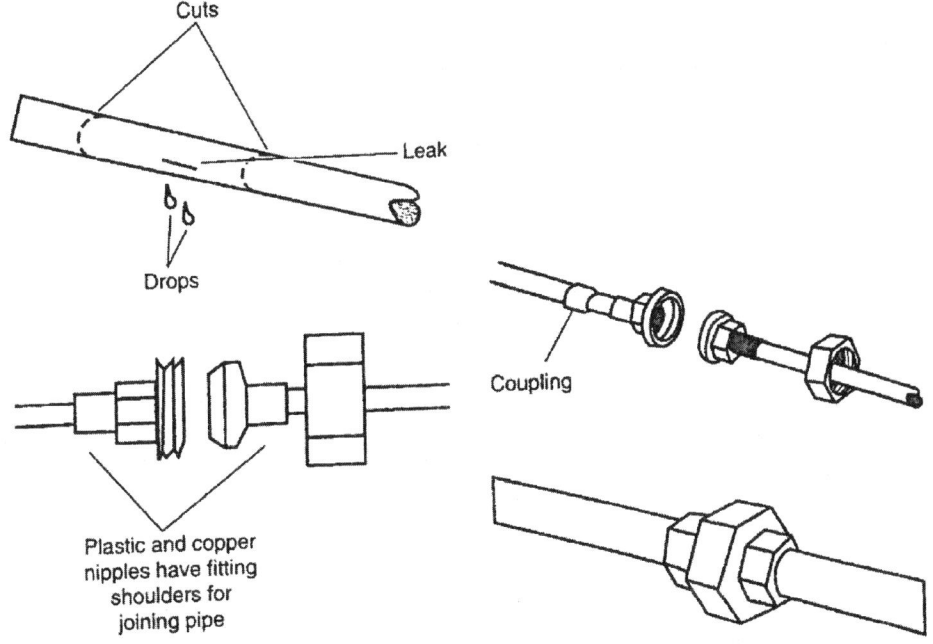

FIGURE 9.10 Unions are made for copper and plastic pipes.

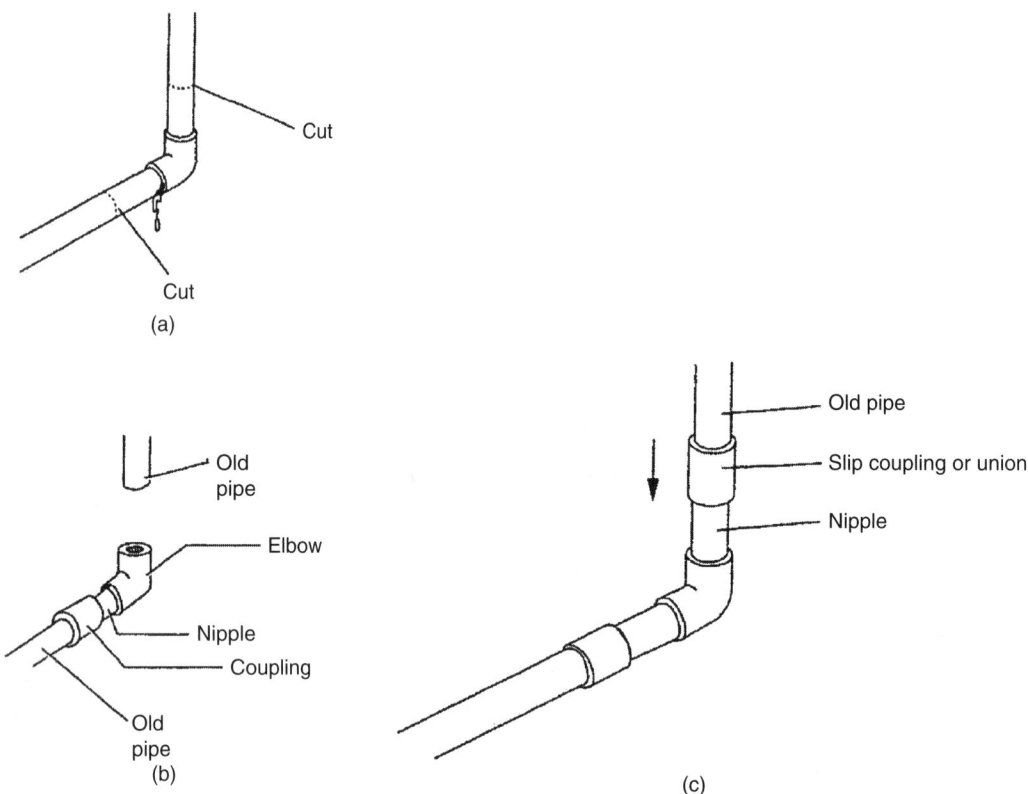

FIGURE 9.11 Fixing a leaky fitting for copper and plastic pipe: (a) Cut out the bad fitting. (b) Build up one side with standard fittings. (c) Use a nipple and union or slip coupling to complete the job.

Compression unions have compression fittings on either side and are typically about 1 inch in length, so you may need two of them for a break of more than about 1 inch. However, compression unions are not regarded as permanent repair in all parts of the country.

To fix a leak in a fitting such as an elbow, you must remove the elbow. While you can sometimes just resolder a copper fitting, it usually does not work. The reason is that in most cases there is just enough water in the elbow somewhere to keep it from getting enough heat on the fitting to fuse the solder. In the long run, it is generally quicker and easier to make the cuts shown in Figure 9.11 and make the repairs as indicated.

Repairing Galvanized Steel Pipe

The basic idea in fixing a leak or a fitting on galvanized pipe is about the same as for other pipe. However, the major difference is due to the way the pipe is assembled by threading. You cannot unscrew a piece of pipe from the middle. To unscrew the pipe, you must start at one end and disassemble back to your trouble point. Of course, this is simply not possible in most situations. The solution is to cut the pipe.

As a rule, you need to cut only one pipe. The rest of the problem area can be unscrewed to fix a pipe leak. Follow the procedure shown in Figure 9.12. First, make a cut an inch or so away from the leak. The handiest tool for this will probably be a reciprocal saw. A pipe cutter will work if there is room, and a hacksaw or mini hacksaw is the next best choice. Next, unscrew the damaged part. If the leak is in the middle of a long pipe, you should probably cut it out.

Use a nipple to replace the damaged area and a union to join the cut sections, as shown in Figure 9.12. However, you must thread the end of the pipe you left in place. If the length of the pipe to the next fitting is visible, you can unscrew the pipe and thread it on a workbench or other convenient place. If not, you can use a pipe die and die wrench to thread it in place. If access is limited, use the ratchet feature of the die wrench and then assemble the repair as shown.

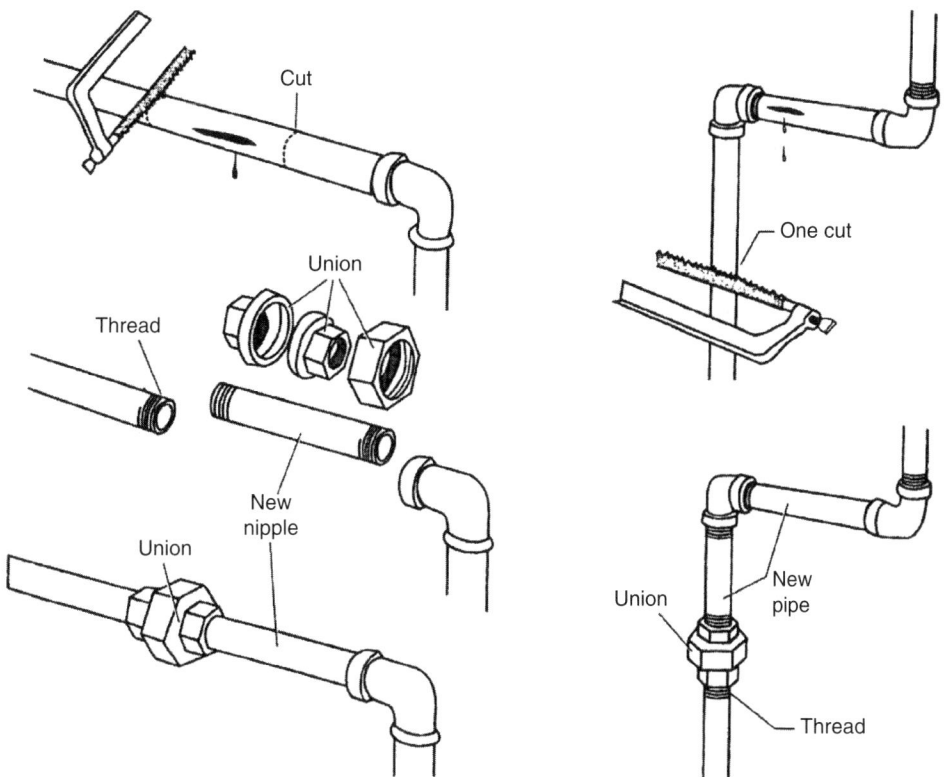

FIGURE. 9.12   Replacing the leaky parts of a galvanized pipe.

If you run into a stubborn fitting that just will not unscrew, loosen it first with a penetrating lubricant and tap it lightly. If that does not work, try heating it with a torch, as shown in Figure 9.13. A minute or so of heat will cause the fitting to expand away from the pipe. Even just a little expansion will likely break loose the thread enough to unscrew. You should remember that the fitting is hot and touch it only with a wrench.

To repair a leaky fitting, usually an elbow, you must again cut one pipe. The best choice is to pick a short area between two fittings, as shown in Figure 9.14. This way, you can use a union and two short nipples for the repair. Cut the shortest length of pipe, as shown in the illustration. Then pull the pipe out just a bit and unscrew either the elbow or the straight side. Next, unscrew the other side. Use the procedure shown in Figure 9.12 to reassemble the pipe and complete the repair.

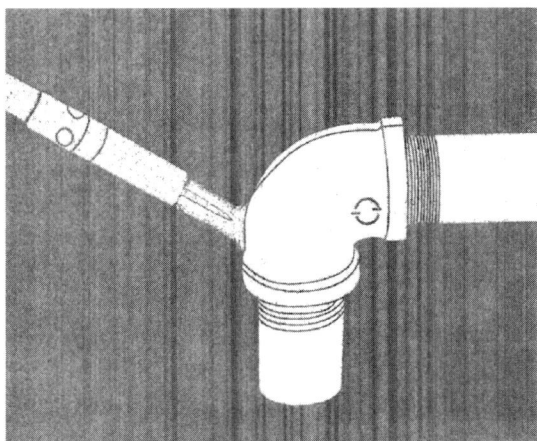

FIGURE 9.13   Heat a stubborn fitting to remove it. Heat only the fitting.

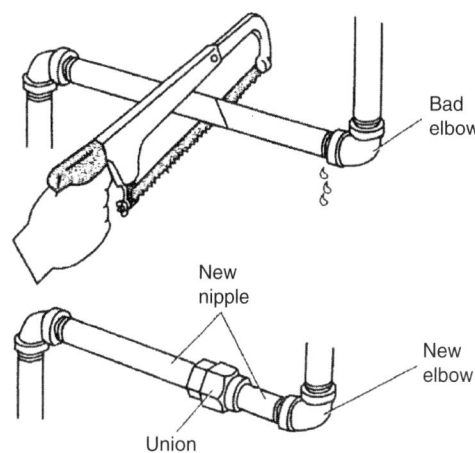

FIGURE 9.14   Fix a leaky galvanized fitting with two nipples and a union.

Making Drain and Vent Repairs

Drain and vent repairs are less common compared with those for supply pipes. However, when you do need to make repairs, the process is the same for either cast iron or plastic. Repairing a failed horizontal pipe is not very difficult. Repairing a failed vertical drain pipe, particularly if it is a cast iron pipe, is usually a job for professionals. That is because the pipe of a vertical drain often supports the weight of the pipe above it, which can be heavy. The weight of the pipe above must be supported before beginning the repair. This may take special tools, particularly for a cast iron main drain on a multistory building.

To repair a horizontal leak, first cut away the damaged area. Before you cut, check and see if you need to support the pipe. Once pipe support is ensured, make the cut an inch or so away from the leak, just as you would for supply pipes. Use the correct diameter of pipe to replace the damaged section. It is an acceptable practice to replace damaged cast iron pipe with PVC sections in almost every code. This is because PVC sections of the pipe are easier to work with.

Complete the repair, as shown in Figure 9.15, by using two flexible couplings. Slide a coupling on each side of the old pipe well past the joint. Then put the PVC repair piece in place and slide the couplings into place on top of it. Tighten the clamp and the repair is complete.

To repair a fitting in a horizontal run, sometimes a rubber or neoprene fitting and clamps can be used. The main consideration is the weight of the pipe. If the weight is allowed to rest on the fitting, it will distort the flexible fitting and cause leaks. If both sections of the pipe are supported so that no weight is on the fitting, the flexible fitting can be used.

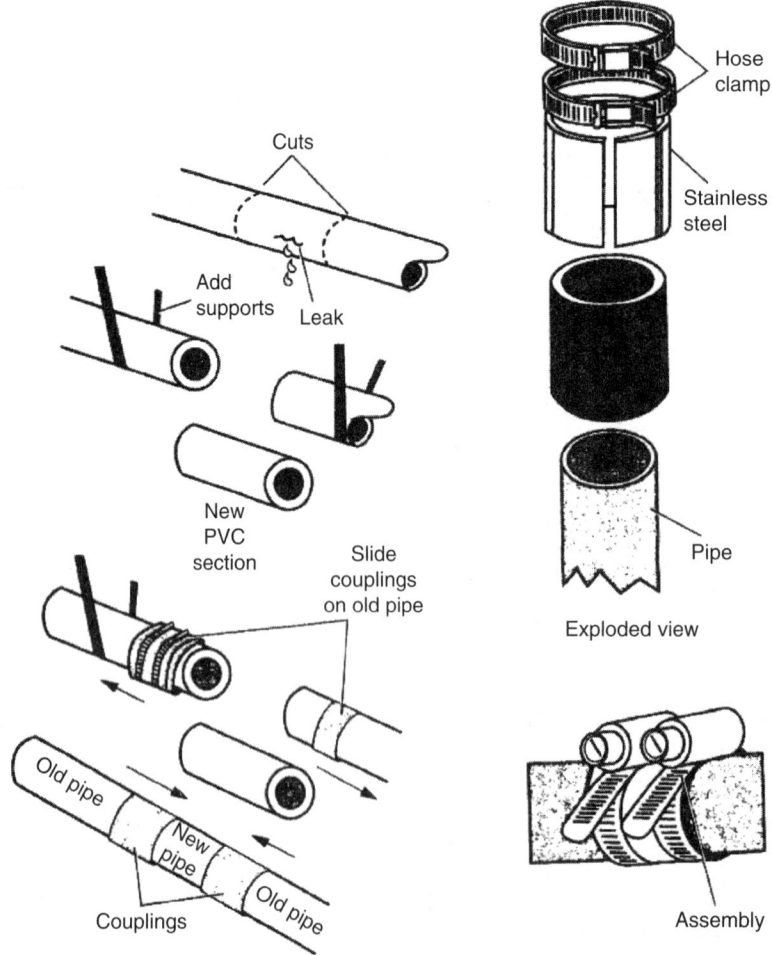

FIGURE 9.15   Replace a bad drain pipe section with PVC and flex couplings.

REVIEW QUESTIONS

1. What does a pipe with a red label and white lettering signify?

   a. flammable fluids and gasses

   b. fire quenching fluids

   c. toxic or corrosive fluids

   d. all water

2. Which color for identifying fluids in pipes can be used in any way by a facility?

   a. purple

   b. pink

   c. tan

   d. none of these

3. How long does it typically take for epoxy to fully cure?

   a. 1 hour

   b. 6 hours

   c. 24 hours

   d. 48 hours

4. What type of material works well for plumbing repairs?

   a. chewing gum

   b. bondo

   c. roll of putty

   d. none of these

5. How do you make a permanent repair in a pipe?

   a. fill the hole with epoxy

   b. use a rubber sleeve

   c. clamp with steel tape

   d. remove the leaky section and replace it with new pipe

6. How do you install slip couplings on copper pipe?

   a. cut the pipe 6 inches on either side

   b. cut the pipe 3 inches on either side

   c. slide the couplings on either side of new pipe and cut enough out to fit the assembly

   d. none of these

7. What can be added to pipe so it can easily be removed at a later time?

   a. nipple

   b. union

   c. elbow

   d. tee

8.   What is a good tool for cutting galvanized pipe?

    a.   handsaw   b.   file

    c.   circular saw   d.   Sawzall

9.   What type of pipe is acceptable to replace damaged cast iron pipe?

    a.   lead   b.   PVC

    c.   copper   d.   none of these

10.   What is used to permanently repair threaded galvanized pipe?

    a.   two nipples and a union

    b.   brazing torch and filler rod

    c.   soldering

    d.   none of these

ANSWERS TO REVIEW QUESTIONS

1. b	2. a	3. c	4. c	5. d
6. c	7. b	8. d	9. b	10. a

Chapter 10
RIGGING

Performance Objectives

After studying this chapter you will (be able to):

1. Explain the differences between certain types of ropes.

2. List the knots and hitches typically used in rigging.

3. Distinguish the various types of shackles and their uses.

4. Understand the proper uses for hooks and eye bolts.

5. Determine when to use different types of beams.

6. Discern the difference between a snatch block and a block and tackle.

7. Identify the various slings and how they are rated.

8. Know the safety aspects concerning cranes.

Rigging is the equipment and the act of planning and installing the equipment in order to move objects. A crew of riggers plan and install the lifting or rolling equipment required to slide or lift objects with ropes, slings, hitches, shackles, hooks, turnbuckles, pipe, and cranes. Rigging is an important aspect of pipefitting, because most industrial/commercial pipe is very large and heavy so it must be moved into place with the help of this equipment to ensure proper alignment.

ROPE

There are two main types of rope that are used for rigging in the construction industry because of their durability and strength. One type is Number One Natural Manila rope that is strong and light yellow in color. As a general rule of thumb, the strength and grade of manila rope decreases as its color darkens. Typically, 1-inch-diameter Number One Natural Manila rope can handle 9000 pounds (4082 kilograms) of weight. The second type of rope is a synthetic rope made of nylon. Since nylon is a plastic, it will not rot, has a higher abrasion resistance, and is more flexible than manila rope. Although it is more expensive than manila rope, a 1-inch-diameter nylon rope can easily handle 22,000 pounds (9979 kilograms) of weight.

DESIGN FACTORS FOR ROPE

When ropes are used to lift objects, they cannot be used to their maximum load capacity. A safety factor for the rope is required because of normal wear and tear, degradation due to exposure to sunlight, excessive angling of load, unknown weight of load, etc. In most cases, a 5:1 design factor is used when lifting loads. A 10:1 factor is typically used when hoisting people or hazardous loads. This means that if the 1-inch-diameter nylon rope could handle a 22,000 pound load, then its load capacity would be divided by either 5 or 10 depending upon the load. Therefore, the 1-inch-diameter nylon rope should lift a load of no more than 4400 pounds if it had a 5:1 design factor and no more than 2200 pounds if it had a 10:1 design factor. This is typically how the working load limit (WLL) is calculated.

HITCHES AND KNOTS

As a rope is tied to a certain length depending upon the size of the load, hitches and knots have to be tied so that it can support the load without unraveling, be tied and untied quickly, maintain rope strength, and ensure the safety of the object being moved as well as the people and equipment moving it. In addition, when a knot or bend is tied in a rope it loses 50% of its original strength. When a hitch is tied in a rope, it loses only 25% of its strength. An eye splice reduces a rope's original strength by 20% and when two ropes are bent in a U shape around each other the rope loses 50% of its strength.

Barrel Hitch

In order to support a barrel vertically, a barrel hitch is tied with a self-centering bowline used to complete the knot as shown in Figure 10.1.

Clove Hitch

A clove hitch is a simple double loop knot that is used to tie a rope over the end of a pipe or post. It can be slipped over the object or tied in position as illustrated in Figure 10.2.

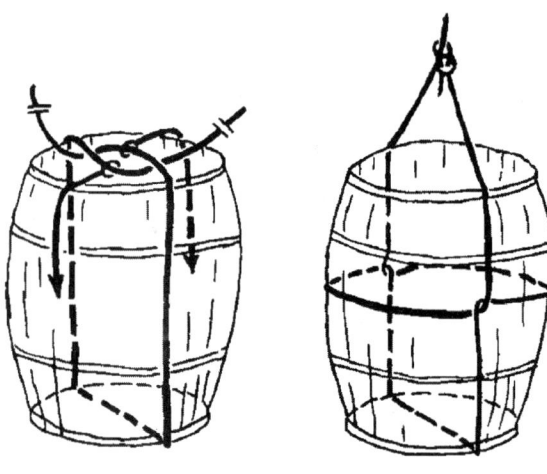

FIGURE 10.1   How to tie a barrel hitch.

FIGURE 10.2   How to tie a clove hitch.

Becket Hitch

A becket hitch is tied so the rope goes around the becket, typically on a set of rope falls as shown in Figure 10.3.

Double Half Hitch

A double half hitch is used to tie off the end of a rope on objects such as posts, rings, and pipes as illustrated in Figure 10.4.

Pipe Hitch

This type of hitch is for what it is named. It is tied to lift a pipe vertically or slide it horizontally as shown in Figure 10.5.

Bowline Knot

This popular knot should never slip or jam under a load as shown in Figure 10.6.

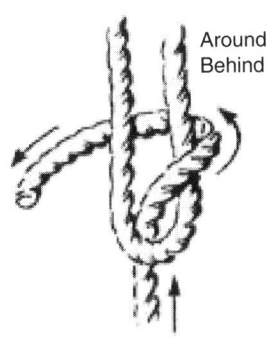

Around
Behind

Up through
From under

Back under itself
Once - single becket bend

Around again back
Under - double becket bend

FIGURE 10.3 How to tie Bechet hitch knots.

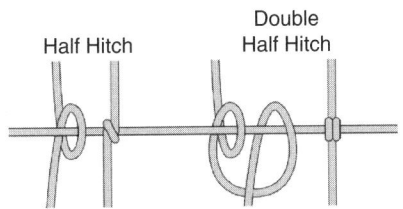

Half Hitch

Double
Half Hitch

FIGURE 10.4 How to tie a double half hitch.

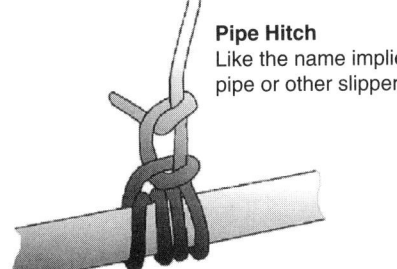

Pipe Hitch
Like the name implies, for pulling on a pipe or other slippery, cylindrical object.

FIGURE 10.5 An example of a pipe hitch.

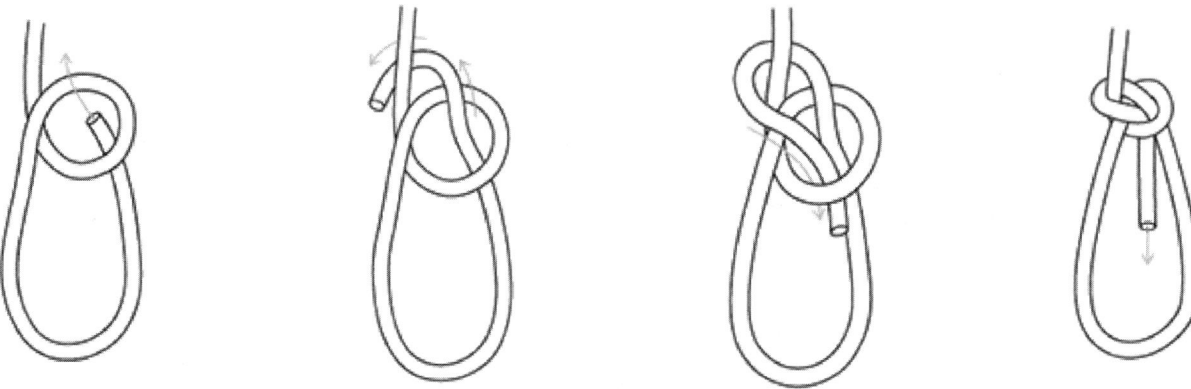

FIGURE 10.6 How to tie a bowline knot.

Bowline on the Bight

This is used to form an eye that should not slip as seen in Figure 10.7.

Self-Centering Bowline Knot

This type of knot is used when a knot must be tied in the middle of a load where equal leg stress must be maintained. Refer to Figure 10.8.

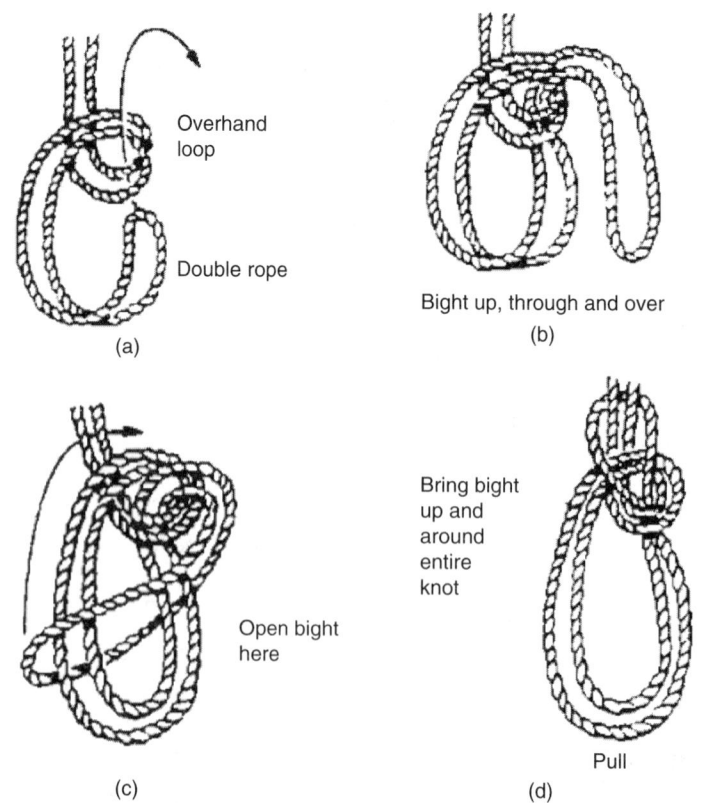

(a) Overhand loop / Double rope

(b) Bight up, through and over

(c) Open bight here

(d) Bring bight up and around entire knot / Pull

FIGURE 10.7 How to tie a bowline on the bight.

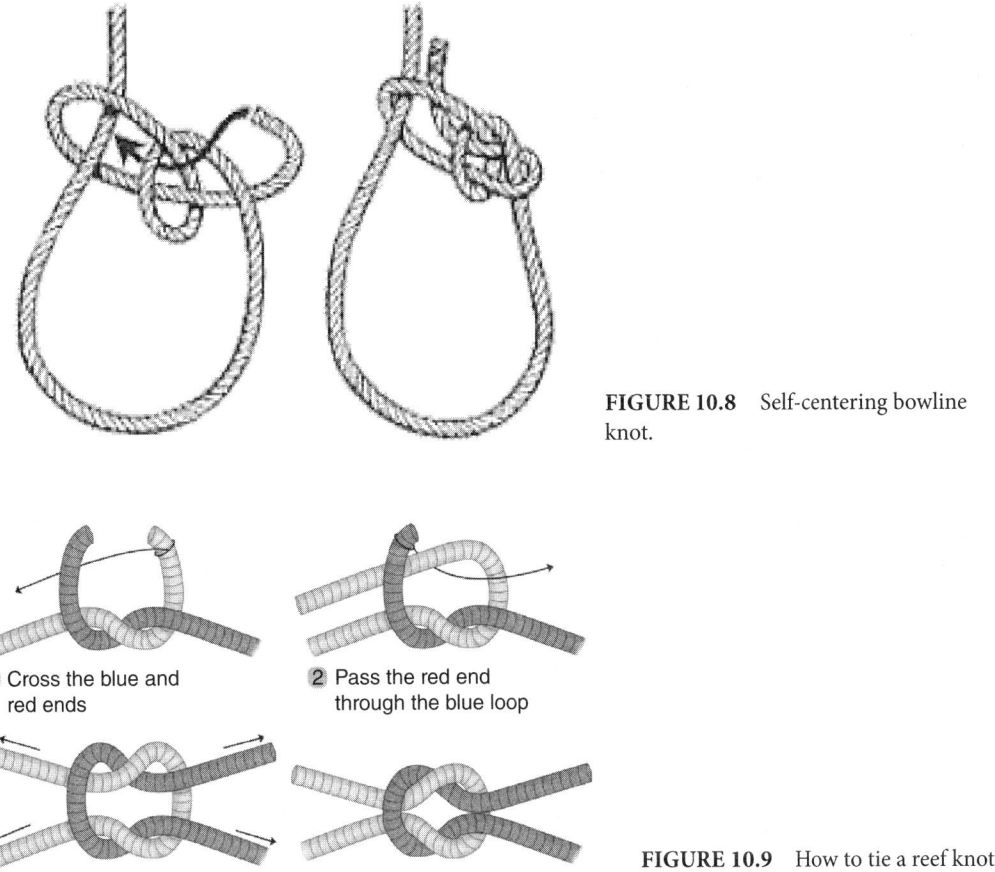

FIGURE 10.8   Self-centering bowline knot.

1 Cross the blue and red ends

2 Pass the red end through the blue loop

3 Pull the ends to tighten

4 The knot is complete

FIGURE 10.9   How to tie a reef knot (101 knots).

Reef Knot

This type of knot is used to attach the two ends of a rope as shown in Figure 10.9. If it is not tied properly, then it can easily come apart, hence the nickname "killer knot."

Cat's Paw

Tying a cat's paw allows a rope to be secured to a hook, especially when it is done with the center of the rope as illustrated in Figure 10.10.

FIGURE 10.10   An example of a cat's paw knot.

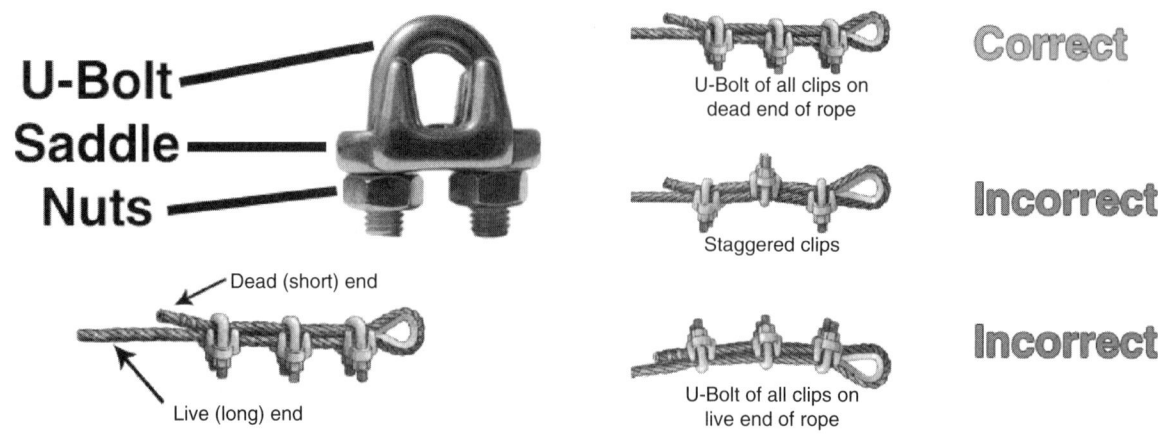

FIGURE 10.11   How to properly fasten U-bolts.

WIRE ROPE

Wires are twisted together to form strands and the strands are then twisted together to form different sizes of wire rope. The center or "core" of the wire rope can be made of fiber or wire, sometimes of a different size than the strands on the perimeter.

Wire Rope Lays

The lay of a rope is important in regard to the use of various types of rope. For instance, Regular Lay has the wire in the strands laid in the same direction while the strands in the rope are laid in opposite directions. This allows the rope to withstand distortion and crushing. In contrast, the wire in Lang Lay is laid in the same direction as the strands. It is used when abrasion resistance is important; however, it is not recommended for single rope hoisting because the strands can unwind. In addition, rope rotation can be prevented under load by Rotation-Resistant ropes that have the inner and outer layers laid in opposite directions (right lay opposing a left lay).

Wire Rope Clips

U bolts are a type of clip that goes over a wire rope that is bent over to create a loop as illustrated in Figure 10.11. This type of clip consists of a U-bolt with threaded ends, a saddle that the U-bolt slips into, and two nuts. It should also be noted that the clip's bottom base (saddle) should be on the live end of the rope and the rounded middle part of the "u" of the bolt should go around the dead end of the rope. All clips should also be facing in the same direction.

   Fist Grip (J-Clip) provides a wider surface matching the contours of the rope for better strength and holding power. In addition, since one nut is placed on the top of the clip and the other one on the bottom, a wrench can easily be rotated for faster installation as shown in Figure 10.12.

   When clips are secured properly, they are equal to approximately 80% of the rated strength of the rope. The number of clips per diameter of wire and length of it turning back are illustrated in Table 10.1.

ANCHOR, CHAIN, AND BOW SHACKLES

Depending on the size and strength of a shackle, it consists of a U-shaped frame, a bolt that slides through or screws into the frame. Furthermore, some shackles will have a nut or pin that is placed on the end of the bolt or for added safety some will have a threaded nut and a cotter pin placed through a hole at the end of

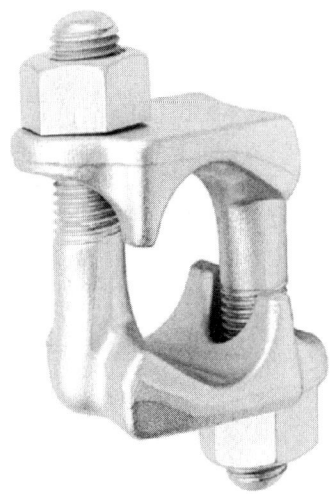

FIGURE 10.12   A picture of a J-clip.

TABLE 10.1   Table listing the number of clips required per rope diameter (Crosby Group).

Clip Size (in.)	Rope Size (in.)	Minimum No. of Clips	Amount of Rope to Turn Back in inches	*Torque in ft lbs.
⅛	⅛	2	3¼	45
3/16	3/16	2	3¾	75
¼	¼	2	4¾	15
5/16	5/16	2	5¼	30
⅜	⅜	2	6½	45
7/16	7/16	2	7	65
½	½	3	11½	65
9/16	9/16	3	12	95
8/8	⅝	3	12	95
¾	¾	4	18	130
⅞	7/8	4	19	225
1	1	5	26	225
1⅛	1⅛	6	34	225
1¼	1¼	7	44	360
1⅜	1⅜	7	44	360
1½	1½	8	54	360
1⅝	1⅝	8	58	430
1¾	1¾	8	61	500
2	2	8	71	750
2¼	2¼	8	73	750
2½	2½	9	84	750
2¾	2¾	10	100	750
3	3	10	106	1200
3½	3½	12	149	1200

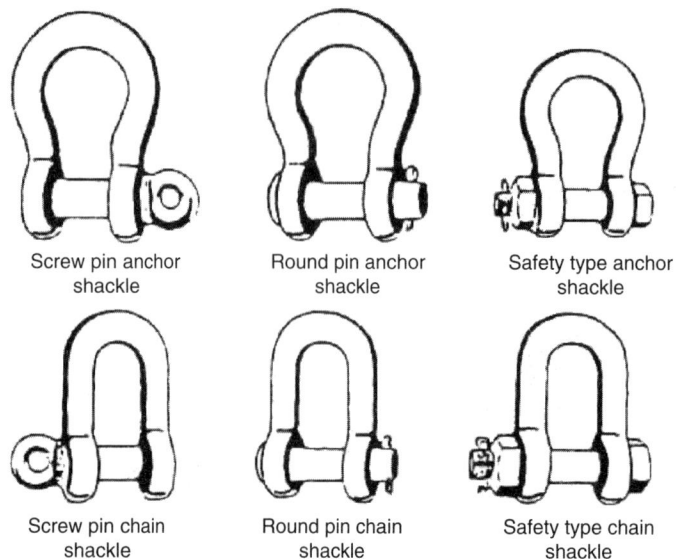

FIGURE 10.13   Various types of shackles.

the bolt. Furthermore, some just have a rounded flat head pin that slips through the U-shaped bow that has a hole in the end to place a cotter pin. As illustrated in Figure 10.13, some are called anchor shackles and have a more rounded shape while others look like pieces of chain and are thereby called chain shackles. Bow-type shackles have a more gradual and wider rounded shape to them.

SHACKLE STRENGTH

The strength of a shackle is based on a straight up pull from the center of the U-shaped bend. If a rope or sling is moved 45 degrees from the top of the shackle, its strength is reduced to 70%. Its strength is reduced to 50% when the rope or sling is at 90 degrees from the top of the shackle. Round pin shackles should not be side-loaded since they only have a round cotter pin holding the horizontal round head pin; refer to Figure 10.14 and Table 10.2.

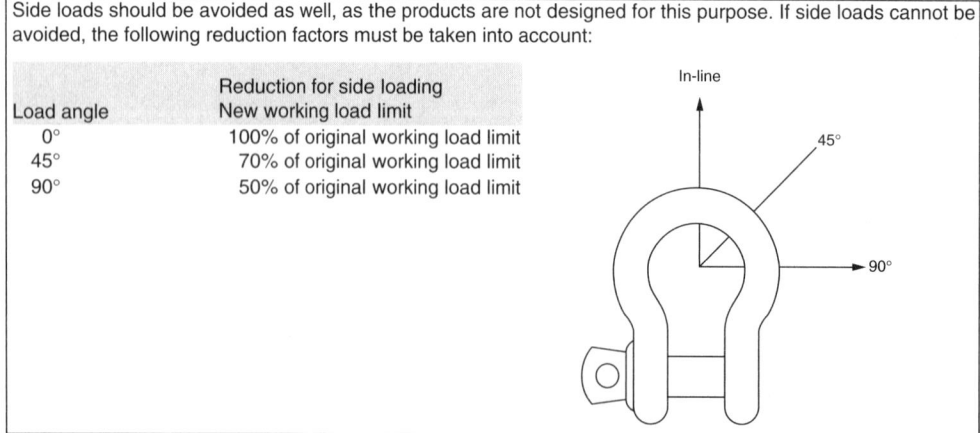

FIGURE 10.14   Shackle strength per load angle.

TABLE 10.2 Shackle strength related to bow size.

Shackles – Quenched and Tempered
Screw Pin and Bolt Type
Design Factor:
Carbon Shackle – 6/1
Alloy Shackle – 5/1

Nominal Size (in.) Diameter of Bow	Carbon Maximum Working Load (tons)	Alloy Maximum Working Load (tons)	Inside Width at Pin (in.)	Diameter of Pin
3/16	⅓	n/a	0.38	0.25
¼	½	n/a	0.47	0.31
5/16	¾	n/a	0.53	0.38
⅜	1	2	0.66	0.44
7/16	1½	2.6	0.75	0.50
½	2	3.3	0.81	0.63
⅝	3¼	5	1.06	0.75
¾	4¾	7	1.25	0.88
⅞	6½	9.5	1.44	1.00
1	8½	12.5	1.69	1.13
1⅛	9½	15	1.81	1.25
1¼	12	18	2.03	1.38
1⅜	13½	21	2.25	1.50
1½	17	30	2.38	1.63

- Ensure screw pin is tight before each lift.
- Use bolt-type shackle for permanent installation.
- Do not side load round pin shackle.
- Use screw pin or bolt type to collect slings.

OVERHEAD LIFTING HOOKS

There are a wide variety of hooks available so it is imperative to use one that is meant for overhead lifting which will have its safe working load limit stamped on it. Always check for cracks, excessive wear, or any type of distortion before using a hook and properly dispose of any that are damaged from use. All hooks should also have a safety catch that is in proper working order. Loads should never be tipped and lifting ropes, slings, etc., should not be placed more than 45 degrees from the center on the hook. In most cases, hooks lifting loads should not be directly connected to the load. Shackles, slings, or other devices should attach to the load which should then be attached to the hook. A hook should not be directly attached to an eye bolt. Refer to Figure 10.15.

EYE BOLTS

Eye bolts can be manufactured with a shoulder or not; refer to Figure 10.16. Non-shouldered (shoulderless) eye bolts are used for straight vertical lifts. Shouldered eye bolts have the extra strength around the neck to be side-loaded. Refer to Table 10.3 and Figure 10.17 for safe working load limits on shouldered eye bolts.

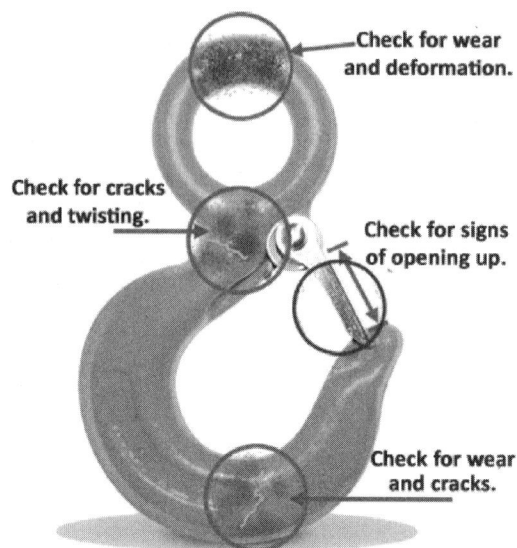

FIGURE 10.15   Hoist and crane depot.

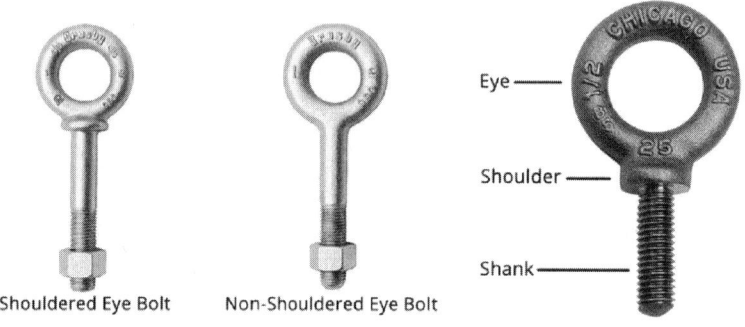

Shouldered Eye Bolt   Non-Shouldered Eye Bolt

FIGURE 10.16   Types of eye bolts.

TABLE 10.3   Working load limits of shoulder eye bolts depending upon angle of pull.

Shank Diameter	Working Load Limit in Line Pull (lbs.)	Working Load Limit 60° Sling Angle (lbs.)	Working Load Limit 45° Sling Angle (lbs.)	Working Load Limit Less than 45° (lbs.)
¼	650	420	195	160
5/16	1200	780	360	300
¾	1550	1000	465	380
½	2600	1690	780	650
⅝	5200	3380	1560	1300
¾	7200	4680	2160	1800
⅞	10,600	6890	3180	2650
1	13,300	8645	3990	3325
1¼	21,000	13,600	6300	5250
1½	24,000	15,600	7200	6000

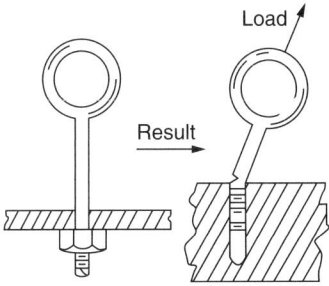

Incorrect use of a shoulderless eye bolt. They do not have the extra rigidity and will bend or break at the shoulder if used for angular loads. Should only be used for straight lifts.

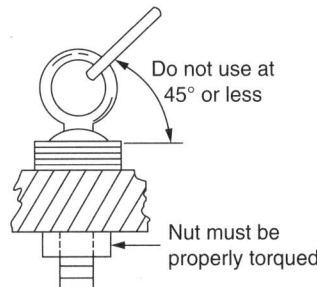

Eye bolt with load correctly applied

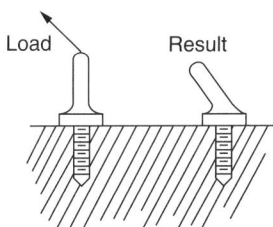

Incorrect way of applying angle load. The ring should be parallel to the direction of the lift, not perpendicular.

FIGURE 10.17   How not to use eye bolts.

BEAMS

Spreader Beam

A spreader beam is used for longer loads that could easily tip or slide while moving. They decrease the possibility of having too long of a sling angle which reduces the load capacity of the sling. Refer to Figure 10.18.

Lifting Beam

A lifting beam is somewhat different than a spreader beam in that it can be directly attached to a hook if vertical space is limited as illustrated in Figure 10.19. Some spreader and lifting beams can be adjusted for unbalanced loads as shown in Figure 10.20.

FIGURE 10.18 A spreader beam (Mazzella Companies).

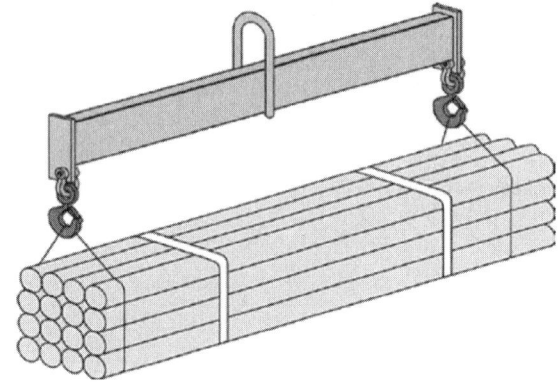

FIGURE 10.19 Lifting beam.

Bail adjusted for
an unbalanced load.

Lift points can be adjusted
for various lengths.

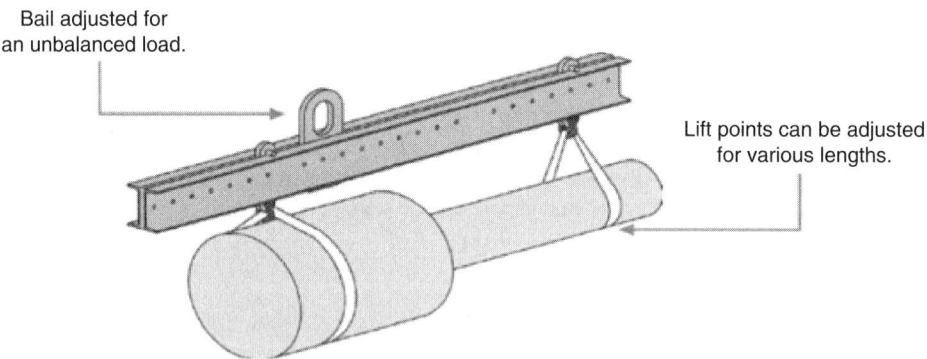

FIGURE 10.20 Adjustable lifting beam.

SHEAVES

Sheaves, pronounced "shivs," are the grooved wheel inside a pulley. The rope used inside a sheave should fit the contour groove of the sheave. It should be noted that the bottom of the groove that a rope rides on should support at least 120-150 degrees as illustrated in Figure 10.21.

If the proper support is not given to the rope in a sheave then premature failure may result. Therefore, the correct sheave should be used with the correct diameter rope.

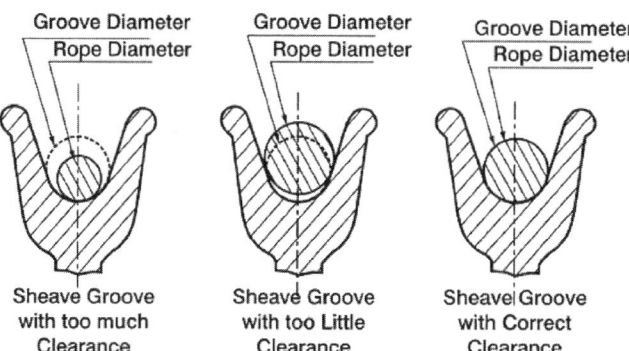

FIGURE 10.21 Side view of sheave illustrated rope diameter and correct clearances.

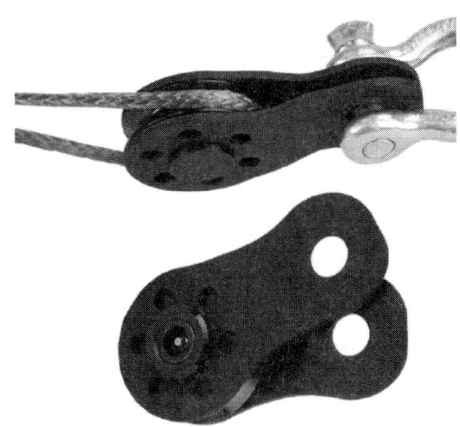

FIGURE 10.22 An example of a snatch block and how it can easily be opened by unscrewing the shackle pin.

SNATCH BLOCKS

In order to change the direction of a rope on a pulley system, a snatch block is used. The block is able to open so rope does not have to feed through as illustrated in Figure 10.22.

BLOCK AND TACKLE

A block and tackle is a combination of pulleys put together to obtain a mechanical advantage so less force has to be used to move a heavy load. Placing pulleys next to each other in a block formation and then adding them underneath in the same fashion allows the ropes to divide the weight of the load by the number of ropes going between each pulley or "sheave." Therefore, if there are two single pulleys placed one above the other, then there would be two ropes dividing the load. If the load being lifted is 100 lbs., then that would be divided by two and only 50 lbs. of force would be required to lift it. As noted in Figure 10.23, depending upon the number of sheaves inside the block, a block and tackle has a different name. These names were originally associated with sailing ships where block and tackles were used to help sailors move heavy masts and extend sails to move with the way in which the wind blew. Industrial blocks and tackles using wire rope are used by pipefitters to move large, heavy commercial pipe.

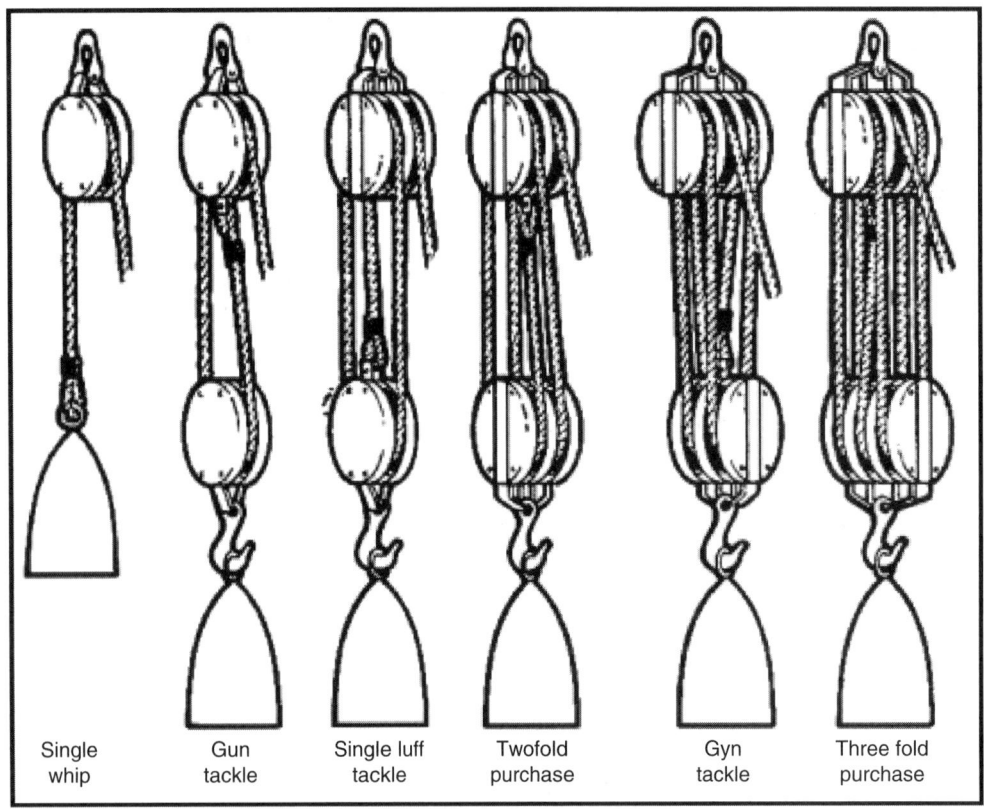

| Single whip | Gun tackle | Single luff tackle | Twofold purchase | Gyn tackle | Three fold purchase |

FIGURE 10.23   Different types of blocks and tackles.

SLINGS

There are a wide variety of slings made in an assortment of materials with different load ratings. Each type of sling is used for certain applications. The types of slings, applications, loads, and their identification are listed in the following sections.

Synthetic Slings

Synthetic slings are frequently used to lift loads because they come in a variety of sizes and load capacities and more importantly are flexible so they do not damage the load. In addition, they are much lighter and less expensive than other types of slings. There are round synthetic slings and flat web woven slings. They are made of nylon or polyester. The polyester slings are about 1/3 to twice as expensive, but are more resistant to acids and other chemicals. In addition, polyester slings only stretch a maximum of 3% when nylon slings can stretch up to 10%. Most synthetic slings are also color-coded dependent upon their lifting capacity. Slings are also identified by number which more or less identifies how the ends are made. The numbering system of the ends is listed as follows:

Type 1—A choker sling has forged alloy end fittings: triangle at one end, choker at the other. The webbing forms a slip noose by slipping the triangle through the choker.

Type 2—A triangle sling has forged alloy end fittings for single or double sling basket hitches. The soft webbing conforms to the load shape without damaging painted or polished surfaces.

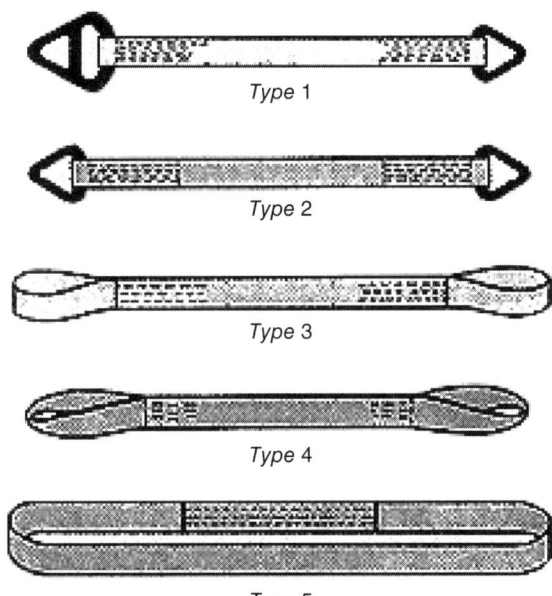

FIGURE 10.24   Illustration of the five number types of slings and their respective ends.

Type 3—A double eye sling is commonly used as a basket and/or choker hitch. For a choker hitch pass, one eye through the other flat eye also makes for easy withdrawal from beneath loads.

Type 4—A reverse eye sling can make a full functional contact with lifting hooks without the use of hardware. This type is widely used for both choker and basket hitches.

Type 5—An endless sling is the most versatile and widely used sling. It is an economical answer to most lifting problems and is ideal for vertical, choker, and basket hitches. Refer to Figure 10.24.

There are some standard color coding of slings in place, however, many manufacturers may use their own color code. For the most part, many manufacturers follow the BS EN-1492 British and European standard as listed below:

1. Purple lifting slings are capable of lifting up to one ton of weight maximum.

2. Green lifting slings are capable of lifting up to two tons of weight maximum.

3. Yellow lifting slings are capable of lifting up to three tons of weight maximum.

4. Gray lifting slings are capable of lifting up to four tons of weight maximum.

5. Red lifting slings are capable of lifting up to five tons of weight maximum.

6. Brown lifting slings are capable of lifting up to six tons of weight maximum.

7. Blue lifting slings are capable of lifting up to eight tons of weight maximum.

8. Orange lifting slings are capable of lifting between 10 and 20 tons, depending on the chosen strength.

Furthermore, every sling has a label sewn on to it which lists a number of things, but most importantly, its load capacity depending upon if it is used in straight vertical lifting, as a choker or as a basket sling as noted in Figure 10.25.

A sling should not be used if the label is missing, is frayed, or if it is damaged in any way. Some of the identification that is listed on a sling is shown in Figure 10.26.

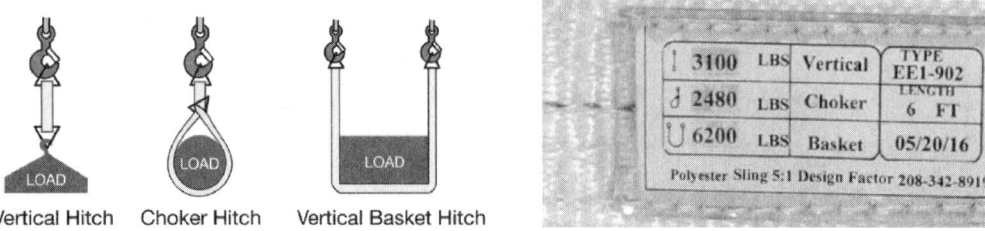

FIGURE 10.25   The three main hitches listed on a label sewn on synthetic slings.

Sling Type:
TC - triangle choker,
TT - triangle triangle,
EE - eye and eye,
EN - endless
    Number of plies: 1, 2, 3, 4
      Webbing grade: 8 or 6
       Sling width (in.)
          Sling Length (in.)
EE 2-9 04 x 12

FIGURE 10.26   Example of round sling identification information listed on the label.

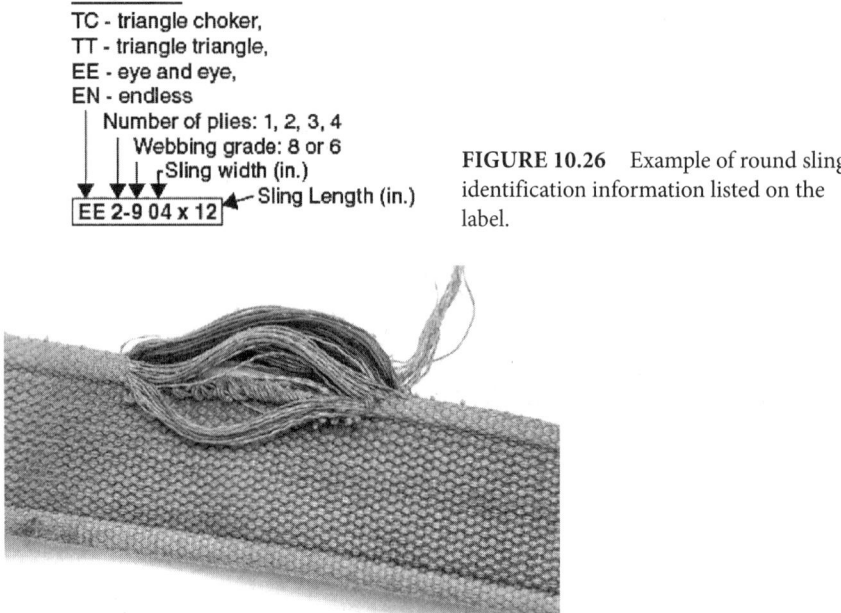

FIGURE 10.27   Damaged sling.

Discard a sling when colored threads are exposed. Structural integrity has been compromised and will fail below the rated load capacity, as illustrated in Figure 10.27.

Wire Rope Sling

Wire rope slings are wire cable slings that are stronger, more durable, and have a higher heat resistance than nylon slings. Typical wire rope slings can withstand temperatures up to 400°F while nylon slings start to degrade after 200°F. The disadvantages of wire rope slings are (1) they are much heavier than synthetic slings, (2) are not flexible or soft, (3) and can damage loads without padding.

Chain Sling

One of the most durable and strong types of rigging available is a chain sling. Many chain slings have been in service for decades. Chain slings offer two important advantages over synthetic slings by being more temperature- and abrasion-resistant while having the ability to be adjusted .

Chain is produced in different grades and the bigger the number, the stronger the chain. Most chain sold in the hardware stores is a 30 or 40 grade. Grades acceptable for lifting heavy loads used for chain slings is a grade 80, 100, and 120 crane. This grade of chain is more apt to stretch than break when overloaded past its rated capacity. Therefore, riggers will have some warning before a failure and be able to replace chains that have been damaged from extended use.

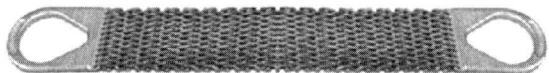

FIGURE 10.28　An example of a metal mesh sling.

Metal Mesh Sling

Metal mesh slings are widely used in metalworking and in other industries where loads are abrasive, hot, or will tend to cut web slings. Unlike nylon and wire rope slings, metal mesh slings resist abrasion and cutting. Metal mesh slings combine alloy steel fittings joined to the steel mesh as shown in Figure 10.28.

CRANE AND HOIST

On occasion, pipefitters will have to weld and fit pipe that may be used in large pipelines and will require the use of a crane to move the very large, heavy load. Since these loads are often quite large and cumbersome, the crane operator needs assistance to make sure he or she will not get too close to power lines and the soil underneath him or her will not shift while moving an object. Therefore, a pipefitter should at least have a rudimentary knowledge of crane and hoist safety so neither the load is damaged nor any workers get injured.

According to OSHA standards, any part of the crane and load should maintain a 20-foot clearance from power lines up to 350 kV and a greater distance from power lines exceeding that voltage. Refer to Table 10.4.

A pipefitter could assist a crane operator by moving to a position where the crane operator cannot see to make sure these limits are maintained. In addition, if a pipefitter knows the crane and hoist signs listed in Figure 10.29 then he or she can assist the operator with directing the load to the correct location without any accidents.

TABLE 10.4　Minimum clearance distances for cranes from power lines.

Voltage (Nominal, kV, Alternating Current)	Minimum Clearance Distance (feet)
up to 50	10
over 50 to 200	15
over 200 to 350	20
over 350 to 500	25
over 500 to 750	35
over 750 to 1000	45
over 1000	(as established by the power line owner/operator or registered professional engineer who is a qualified person with respect to electrical power transmission and distribution)

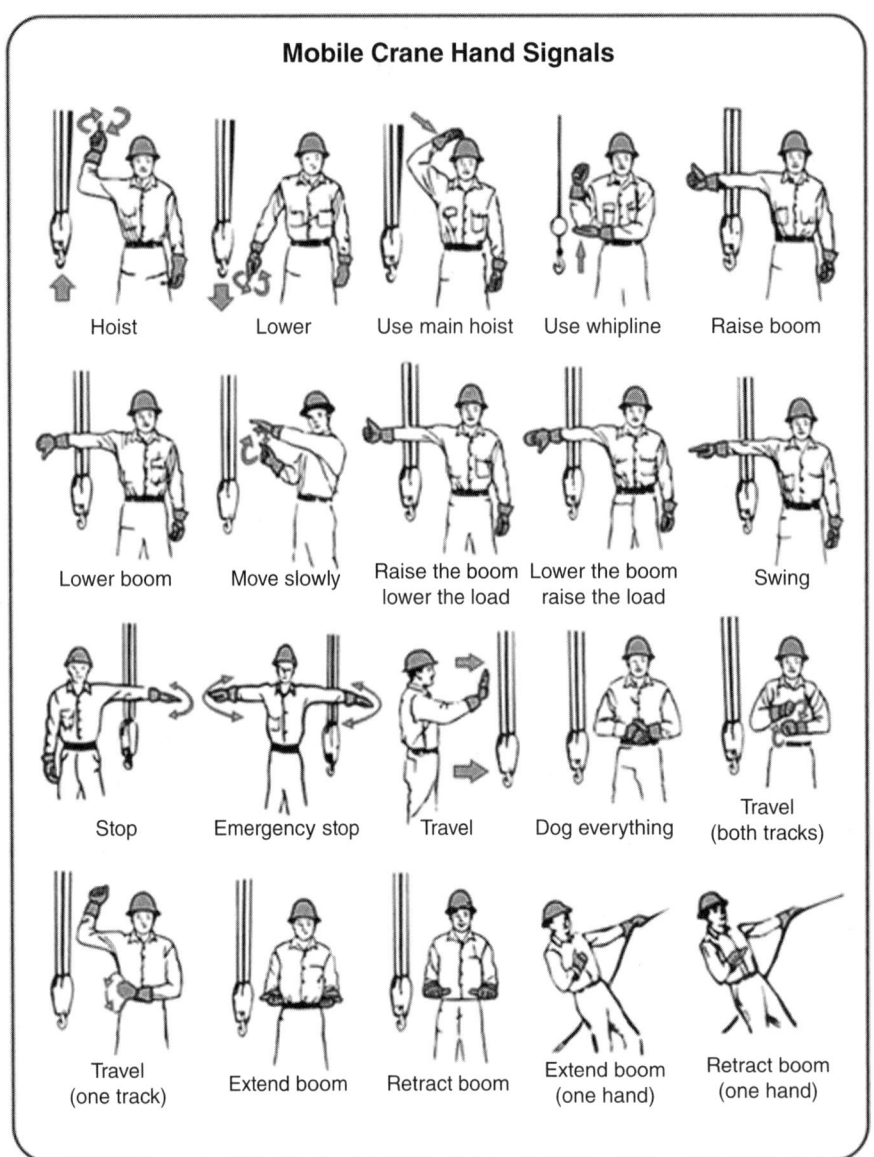

FIGURE 10.29   Mobile crane hand signals (Stevenson Crane Service, Inc.).

REVIEW QUESTIONS

1. As a general rule of thumb, the grade of rope decreases as its color _____.

   a. lightens
   b. darkens
   c. fades
   d. turns yellow

2. Natural rope that is used for rigging is typically made from

   a. nylon
   b. bamboo
   c. straw
   d. manila

3. A 1-inch-diameter nylon rope has _____ the load capacity of a similar diameter natural rope.

   a. twice
   b. triple
   c. half
   d. a quarter

4. What is the typical design factor for rope when hoisting people or hazardous loads?

   a. 4:1
   b. 5:1
   c. 6:1
   d. 10:1

5. What happens when a hitch is tied in a rope?

   a. the rope loses 20% of its strength
   b. the rope loses 25% of its strength
   c. the rope loses 50% of its strength
   d. the rope strengthens by 25%

6. What is the name of a hitch that is a simple double loop knot that is used to tie a rope to a pipe?

   a. barrel hitch
   b. becket hitch
   c. clove hitch
   d. double half hitch

7. Which type of knot is also called the "killer knot" because of the way it is tied?

   a. bowline
   b. bowline on the bight
   c. cat's paw
   d. reef

8.   What could be done to prevent rope rotation when moving loads?

   a.   place inner and outer layers in the same direction

   b.   place inner and outer layers in opposite directions

   c.   place each strand of the rope in the same direction

   d.   alternate the strands of the rope in opposite directions

9.   How should U-bolts be placed when securing the end of a rope?

   a.   staggered

   b.   in pairs

   c.   all U-bolts should clip on the live end side of the rope

   d.   all U-bolts should clip on the dead end side of the rope

10.   What is the advantage of a J-clip?

   a.   when secured properly, they are rated at 50% of the rope strength

   b.   when secured properly, they are rated at 100% of the rope strength

   c.   only one fastening nut is put on the top and on the bottom for ease of tightening

   d.   the nuts snap in with minimal force

11.   What do some shackles have for added safety and strength?

   a.   nuts that are welded to the ends

   b.   nylon insert locking nuts

   c.   a nut on each end with a cotter pin that is placed in a hole after the nut

   d.   none of these

12.   Which type of shackle has a donut-shaped pin?

   a.   round pin anchor   b.   flat pin anchor

   c.   screw pin anchor   d.   safety type anchor

13.   If a sling is moved 45 degrees from the top of the shackle, its strength is reduced to

   a.   90%   b.   70%

   c.   50%   d.   25%

14.   Loads on a hook should never be placed _____degrees from the center.

   a.   25   b.   45

   c.   60   d.   30

15. A hook should not be directly attached to

    a. a sling
    b. a shackle
    c. a rope
    d. an eye bolt

16. What is used to strengthen an eye bolt?

    a. tapered neck
    b. webbing
    c. shoulder
    d. offset

17. What can be used to reduce the sling angle when lifting a load?

    a. spreader beam
    b. chain
    c. wire rope
    d. sheaves

18. Which type of beam can be directly attached to a hook if vertical space is limited?

    a. spreader
    b. lifting
    c. sheaves
    d. hoist

19. How many degrees of a rope should be supported by a sheave?

    a. 45
    b. 60
    c. 90
    d. 120

20. How would the direction of a rope on a pulley system be changed?

    a. snatch shackle
    b. block pin
    c. hitch block
    d. snatch block

21. If a block and tackle has three pulleys on the top and bottom, how much force is required to lift a 600-pound object?

    a. 50
    b. 100
    c. 200
    d. 300

22. A choker sling has what type of ends?

    a. a forged alloy steel triangle on each end
    b. an eye on either end
    c. a reverse eye on either end
    d. a forged alloy steel triangle on one end and a similar forged steel triangle with a slot in it

23. Typically, a yellow sling is capable of lifting

a. 1 ton   b. 2 tons

c. 3 tons   d. 4 tons

24. What does the 2 in the following sling label represent? EE 2-9 04 × 12

a. width of the sling   b. length of the sling

c. type of sling   d. number of plies in the sling

25. Which type of slings is resistant to heat, abrasion, and cutting and can be made in various widths?

a. synthetic   b. chain

c. wire rope   d. metal mesh

ANSWERS TO REVIEW QUESTIONS

1. b	2. d	3. a	4. d	5. b
6. c	7. d	8. b	9. d	10. c
11. c	12. c	13. b	14. b	15. d
16. c	17. a	18. b	19. d	20. d
21. b	22. d	23. c	24. d	25. d

Chapter 11
PIPE OFFSETS

Performance Objectives

After studying this chapter you will (be able to):

1. Understand the importance and purpose of pipe offsets.

2. Know the terminology associated with pipe offsets.

3. Demonstrate how to use formulas and constants to calculate pipe offsets.

4. Explain the importance and purpose of fitting allowances.

5. Define the terminology associated with fitting allowances.

6. Demonstrate how to use formulas and constants to calculate fitting allowances.

7. Understand the importance and purpose of spread between pipes.

8. Know the terminology associated with spread.

9. Demonstrate how to use formulas and constants to calculate spread and spread allowances.

10. Demonstrate how to use formulas and constants to calculate rolling offsets.

Pipes are different than tubing because they cannot be bent to move around an obstacle. Fittings, such as elbows, tees, couplings, etc., are used to change the direction of solid, inflexible pipe. Unlike smaller diameter residential plumbing, pipefitters must work with large diameter pipe that must be either welded in place or threaded where exact dimensions are of utmost importance. Due to these concerns, pipefitters must be able to calculate the length of pipe that would be diverted at set angles created by fittings. In other words, if the pipe is diverted or "offset" by a 45-degree fitting to bypass an obstacle, then they would have to know how to calculate the length of the new angled pipe. Certain terminology is used to identify these angles as illustrated in Figure 11.1.

CALCULATING PIPE OFFSETS

Travel is the angled distance that connects from the starting point of the pipe to where it is now "traveling." Typically, it would be the hypotenuse of the triangle that is formed when the pipe is diverted or the longest side.

Offset or sometimes referred to as "set" is how far the pipe is diverted from its original course.

Run is the distance the pipe would travel in a straight line if it has not been diverted. Run is also referred to as the *advance*.

When working with angles and the triangles they form, it is best to apply the Pythagorean theorem and trigonometry. In order to save time in calculating the sine, cosine, and secant for all the various angles

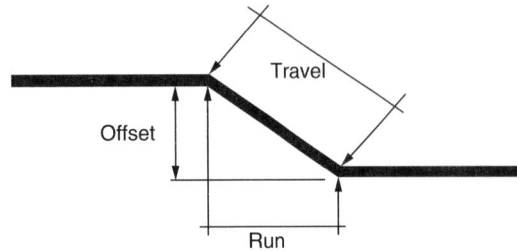

FIGURE 11.1   Pipe offset terminology.

TABLE 11.1 Formulas and constants for calculating standard pipe offsets.

Formula	72-degree Elbow	60-degree Elbow	45-degree Elbow	30-degree Elbow	22½-degree Elbow	11¼-degree Elbow	5⅝-degree Elbow
Travel = Offset ×	1.052	1.155	1.414	2	2.613	5.126	10.187
Travel = Run ×	3.236	2	1.414	1.155	1.082	1.019	1.004
Run = Offset ×	0.325	0.577	1	1.732	2.414	5.072	10.158
Run = Travel ×	0.309	0.5	0.707	0.866	0.924	0.98	0.995
Offset = Run ×	3.078	1.732	1	0.577	0.414	0.198	0.098
Offset = Travel ×	0.951	0.866	0.707	0.5	0.383	0.195	0.095

created by the fittings. Table 11.1 lists the formulas with constants for the standard pipe elbows used to divert pipe. Using the numbers in Table 11.1, most of the calculations for making pipe offsets can easily be determined.

For example, if a pipefitter was using a 45-degree elbow to offset the line 12" then how long would the travel (angled pipe) have to be cut to?

1. Travel = offset × 45-degree elbow constant from table

2. Travel = 12" × 1.414

3. Travel = 16.968"

NOTE: All calculations are inferring that distances are taken from the center of a pipe and elbow.

FITTING ALLOWANCE

Once the travel length is calculated from the center of the two fittings of offset parallel pipe lines, the length will be longer than what is really required to fit inside the fitting. This is because the pipe does not need to fit too far into a fitting to make a good seal and/or the pipe would hit the part of the fitting where it is curving into another direction before it reaches the center line. Moreover, the additional length of pipe in the fitting would be a waste of material and increases the cost of the pipe. Therefore, depending upon the joint used to connect the pipe to the fitting, some of the length of the travel would have to be subtracted which is termed the "fitting allowance." The fitting allowance will differ depending on the type of pipe and joint made to connect it. For instance, a threaded pipe will have to screw into a fitting while a large diameter steel pipe may be butt-welded to a fitting. At any rate, a formula for a threaded pipe that has to be inserted into a fitting would depend upon the following terms:

Center-to-center (CC) is the distance from the center of one fitting to the center of the fitting on the other end of the pipe also known as the *travel* distance of the pipe.

Face-to-center (FC) is the area from the front face of one of the openings for a pipe fitting to the center point of the other end of the fitting. FC is also referred to as *fitting dimension*.

Thread engagement (TE) is the threaded part inside the fitting that the pipe threads into. TE is also referred to as the *socket depth*.

Fitting allowance (FA) is the distance left over when the thread engagement is subtracted from the FC.

End-to-end (EE) is the actual length of pipe that would fit properly in the threaded fittings.

Therefore, in order to obtain the actual length of the threaded pipe that should go in between the two threaded pipe fittings, the thread engagement would have to be subtracted from the FC distance to obtain the additional fitting allowance that is not needed as illustrated in Figure 11.2. In other words, FA = FC − TE. Since the pipe is threaded on both sides to fit in the threaded fittings, the FA would be subtracted twice from the CC or travel distance that was calculated previously. The answer would be the EE or actual length of the threaded pipe that will fit between the two fittings as illustrated as hidden (dashed) line in Figure 11.3.

For example, if a pipe fitting's FC distance was 4 inches and the thread engagement was 1¼ inches then how long would the actual pipe be cut if the CC of the two fittings was a distance of 36 inches?

1. First find the fitting allowance which is: FA = FC − TE.

2. FA = 4" − 1¼"

3. FA = 2¾"

4. EE = CC − 2 × FA

5. EE = 36" − 2(2.75")

6. EE = 36" − 5.5"

7. EE = 30.5" or 30½"

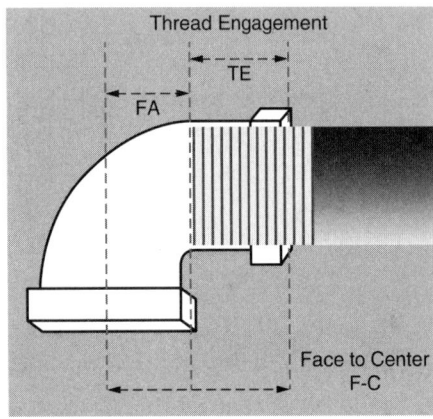

FIGURE 11.2 Diagram of fitting allowance terminology.

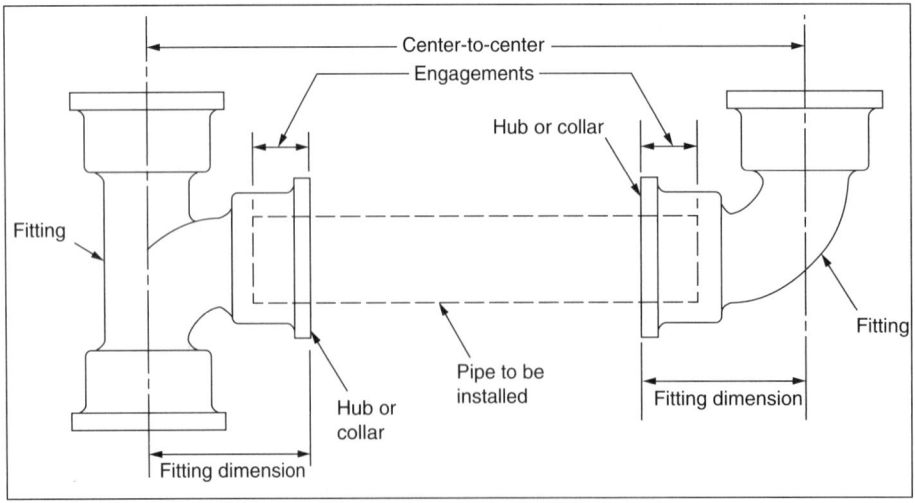

FIGURE 11.3 Determining actual pipe length diagram.

PARALLEL AND MULTIPLE OFFSETS

When more than one pipe is diverted at some specified angle for any reason, then multiple pipe offsets are created. In most cases, the pipes should maintain an equal distance from each other which is known as the *spread (S)*. This distance is maintained for a number of reasons. First of all, it is more aesthetically appealing when all the pipes follow each other at equal distances from each other. Secondly, they balance a system which is important when pipes are used to carry hot and ice-cold water for HVAC systems. Unbalanced systems make a building creek and groan when the systems are turned on and off, and in some cases could cause a boiler to overheat and explode. Lastly, if the pipes are suspended from the ceiling and are at equal distance from each other, then it is easier to make a rack that fits them.

The issue with multiple pipes maintaining an equal distance is that they cannot be bent at the same place and be expected to align with each other as illustrated in Figure 11.4. If this is done, you will notice that the spacing where the pipes are diverted at an angle becomes closer. Therefore, the fittings must be placed farther ahead from the previous fitting to counteract this effect. This is referred to as the *spread allowance (SA)* as shown in Figure 11.4.

CALCULATING SPREAD ALLOWANCES

As illustrated in Figure 11.5, the SA is the extra distance that the second pipe needs so the pipes can maintain an equal spread throughout the bend and back. In order to calculate the spread, you would use trigonometry; however, Table 11.2 has the formulas and constants already calculated for the basic offset angles made by pipefitters.

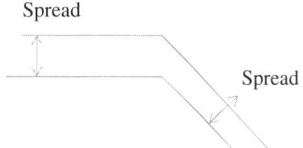

FIGURE 11.4 Illustration of unequal spread of pipes from each other due to straight alignment of fittings. Note: the lines representing parallel lines.

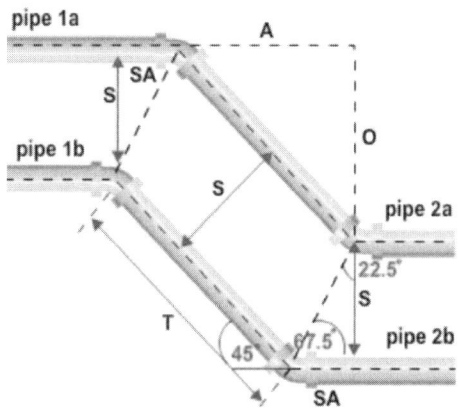

FIGURE 11.5 Pipe diagram of spread (S) and spread allowance (SA).

TABLE 11.2 Formula and constants for calculating spread allowances to maintain equal spread between pipes.

Formula	Offset Angles						
Spread allowance = Spread ×	90	72	60	45	30	22½	11¼
	1	0.727	0.577	0.414	0.268	0.199	0.098

For example, if a pipe is diverted at a 45-degree offset of 10 inches and maintains a spread of 12 inches then what would be the SA for the second pipe.

Answer:

1. SA = Spread × constant

2. SA = 12" × 0.414

3. SA = 4.968"

If there is more than one pipe running parallel to the first pipe then the same SA would be added to the previous pipe's SA, thereby increasing the SA for each and every subsequent pipe. This would then ensure an equal spread between all the pipes.

MAINTAINING SPREAD WITH A 90-DEGREE TURN

Typically, two 45-degree fittings are used to change the direction of a pipeline verses using one 90-degree fitting that would restrict the flow. If a parallel pipe follows the first pipe, then its travel distance would have to be longer than the first pipe to maintain an equal spread.

An example of how to calculate this would be as follows:

Find the travel distances and SA for two pipes that have a 10-inch spread and a 12-inch offset at a 45-degree angle. Refer to Tables 11.1 and 11.2 to assist with the problem.

1. Travel distance for the first pipe = Constant × offset.

2. Travel distance for the first pipe = 1.414 × offset.

3. Travel distance for first pipe = 1.414 × 12" = 16.968"

4. Travel distance for the second pipe = previous travel distance plus the spread allowance × 2.

5. SA = Spread × constant

6. SA = 10" × 0.414

7. SA = 4.14"

8. 16.968" + 4.14" + 4.14"

9. Travel distance for second pipe = 25.248"

ROLLING OFFSETS

Rolling offsets are required to be calculated when a pipe not only changes direction to the left or right, but it also angles up or down simultaneously. This is illustrated in Figure 11.6.

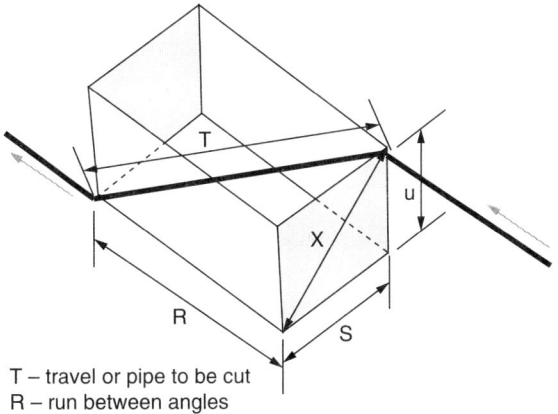

T – travel or pipe to be cut
R – run between angles
S – horizontal offset
U – vertical offset

FIGURE 11.6   Diagram of a rolling offset.

TABLE 11.3   Formulas and constants for calculating common rolling offsets for pipes.

Formula	Offset Angles						
	90	72	60	45	30	22½	11¼
Travel = Offset ×	1	1.052	1.155	1.414	2	2.613	5.126
Run = Offset ×	0	0.325	0.577	1	1.732	2.414	5.027

For example, if the travel angle in Figure 11.6 is 45 degrees and the roll is 12 inches and the rise is 16 inches then what would be the run and the travel?

In order to find the run and travel, the offset must first be calculated in order to use Table 11.3 with common elbow fitting sizes. The formula for calculating the offset in this instance is:

1. Offset $= \sqrt{\text{Roll}^2 + \text{Rise}^2}$

2. Therefore, offset $= \sqrt{12^2 + 16^2}$

3. Offset $= \sqrt{144 + 256}$

4. Offset $= \sqrt{400}$

5. Offset $= 20$

6. Use Table 11.3 for calculating common rolling offsets for pipes to find the travel and run.

7. Travel = Offset × Constant

8. Travel = 20" × 1.414

9. Travel = 28.28"

10. Run = Offset × 1

11. Run = 20" × 1

12. Run = 20"

A framing square, although not as accurate, could also be used to find the offset by laying a straight edge measuring instrument between the lengths of the rise and run as illustrated in Figure 11.7.

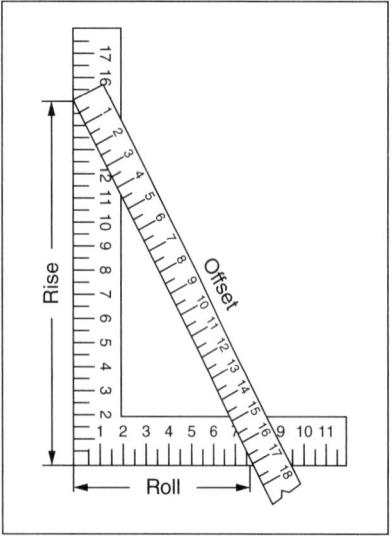

FIGURE 11.7   Finding the offset using a framing square.

REVIEW QUESTIONS

1. What is the name of the angled pipe distance that diverts a pipe from an obstacle?

   a. run

   b. rise

   c. offset

   d. travel

2. The advance is also referred to as the _____.

   a. run

   b. rise

   c. offset

   d. travel

3. What is the term for the triangle's hypotenuse when calculating pipe offset distances?

   a. run

   b. rise

   c. offset

   d. travel

4. The socket depth in regard to a fitting is also referred to as _____.

   a. end-to-end

   b. face-to-center

   c. fitting allowance

   d. thread engagement

5. What is the distance that is leftover when the thread engagement is subtracted from a fittings face-to-center?

   a. end-to-end

   b. center-to-center

   c. fitting allowance

   d. thread engagement

6. Another term for fitting dimension is _____.

   a. end-to-end

   b. face-to-center

   c. fitting allowance

   d. thread engagement

7. What is the term for the actual length of the pipe cut to fit the offset?

   a. end-to-end

   b. face-to-center

   c. fitting allowance

   d. thread engagement

8. What is the term that refers to the distance from the center of one pipe to the center of the next pipe?

   a. offset

   b. fitting allowance

   c. spread

   d. spread allowance

9. What is the term for staggering the place of fittings to divert pipe at equal distances?

    a. offset   b. fitting allowance

    c. spread   d. spread allowance

10. What is the term for a pipe that diverts to the left and rise up several inches?

    a. fitting allowance   b. rise

    c. offset   d. rolling offset

11. If a pipefitter was using a 22½-degree elbow to offset the line 12 inches then how long would the travel (angled pipe) have to be cut to?

    a. 31.356"   b. 12.984"

    c. 28.968"   d. 11.088"

12. If a pipefitter was using a 60-degree elbow to offset the line 4 inches, then how long would the travel (angled pipe) have to be cut to?

    a. 2"   b. 2.308"

    c. 4.62"   d. 8"

13. If a pipefitter was using a 30-degree elbow to offset a line 10 inches then how long is the run?

    a. 20"   b. 17.32"

    c. 11.55"   d. 8.66"

14. What would be the offset of a pipe diverted 30 degrees if the run is 20 inches?

    a. 20"   b. 8.66"

    c. 5.77"   d. 5"

15. If the travel is 22 inches what would be the offset of a pipe if a 30-degree fitting is used?

    a. 20"   b. 8.66"

    c. 5.77"   d. 5"

16. A pipe fitting's face-to-center distance is 3 inches and the thread engagement is 1½ inches so how long would the actual pipe be cut to if the center-to-center of the two fittings is a distance of 24 inches?

    a. 18½"   b. 21"

    c. 22½"   d. none of these

17. A pipe fitting's face-to-center distance is 6 inches and the thread engagement is 1¼ inches so how long would the actual pipe be cut to if the center-to-center of the two fittings is a distance of 64 inches?

    a. 54½"

    b. 55½"

    c. 56¾"

    d. none of these

18. If a pipe is diverted at a 45-degree offset of 4 inches and maintains a spread of 2 inches then what is the spread allowance for the second pipe.

    a. 0.828"

    b. 1.655"

    c. 2"

    d. 2.484"

19. If a pipe is diverted at a 11¼-degree offset of 14 inches and maintains a spread of 4 inches then what is the spread allowance for the second pipe?

    a. 0.392"

    b. 1.1025

    c. 1.372

    d. none of these

20. If the travel angle is 60 degrees and the roll is 12 inches and the rise is 16 inches then what would be the run and the travel distances?

    a. R = 23.1, T = 11.54

    b. R = 11.54, T = 23.1

    c. R = 3.05, T = 6.11

    d. none of these

ANSWERS TO REVIEW QUESTIONS

1. d	2. a	3. d	4. d	5. c
6. b	7. a	8. c	9. d	10. d
11. a	12. c	13. b	14. c	15. d
16. b	17. a	18. a	19. a	20. b

—NOTES—

Chapter 12
READING WELDING BLUEPRINTS

Performance Objectives

After studying this chapter you will (be able to):

1.   List the five main types of welding joints.

2.   Identify and explain when to use certain welds.

3.   Describe various ways to prepare a welding joint.

4.   Distinguish between the main types of groove welds.

5.   Understand the elements of a welding symbol.

6.   Read a welding symbol.

The bulk of a pipefitter's duties, in most cases, is to weld pipe according to plans that are provided by the engineers who designed the system. There are a wide variety of ways to fasten pipe using different joints, welds, penetration, welding process, welding rod, appearance of weld, length of weld, and more. In order to remove the ambiguity from these specifications, certain welding symbols were created to specifically address how a weld was intended by those who designed the piping system. This chapter will review the nomenclature that is involved with welding and what would be listed on plans that a pipefitter would be given to fabricate pipe systems.

WELDING JOINTS

There are five basic welding joints by which metal could be fitted up in order to be welded as shown in Figure 12.1.

1.   *Butt joint* is one of the simplest of joints where the two edges of the pieces to be welded are pushed together and then given a gap of 1/16 to 1/8 inch if the job requires deep penetration for a stronger weld. Most pipe systems require a butt-weld to fasten pipes, fittings, or flanges together.

2.   *Tee-joint* forms the shape of the capital letter T either right side up or upside down. Tee-joints require the welder to make a weld at a 45-degree angle which is called a *fillet weld*. These joints can be welded from one side or both sides depending upon the strength requirements of the joint.

3.   *Lap joint* is when one piece of material is placed on top of another piece of material for a specified amount of length. In this way, a fillet weld could be welded on either side or both sides of where the pieces overlap. Moreover, if holes or slots are cut into one of the pieces, welds could be made inside the edges of these perforations. Furthermore, pieces that are lapped over each other make it easier to resistance weld or spot weld thin pieces of metal. It should also be noted that since the metal is lapped over the other piece of metal, lap joints are considered one of the strongest joints.

4.   *Corner joint* is when one edge is aligned 90 degrees of another edge and then welded down the outside, inside, or both edges. This is considered the weakest of the joints.

5.   *Edge joint* is when two edges are placed next to each other to weld. However, one of the two pieces is usually bent out at an angle so it is actuality forming a corner joint, but with two edges. This makes for a stronger joint than the corner joint.

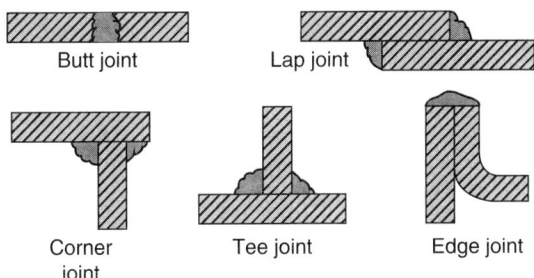
Butt joint   Lap joint

Corner joint   Tee joint   Edge joint

FIGURE 12.1 The five main welding joints used to fabricate parts.

TYPES OF WELDS

There are a wide variety of welds available; however, their use is mainly dependent upon the thickness of the materials to be fabricated as illustrated in Table 12.1. The names of these welds are based upon how the metal edges are prepared before welding. For example, if one edge of the two pieces of material being welded together is beveled at a specific angle then it is called a single-bevel joint.

It should be noted that when the edges of two metal pieces are placed together with no surface preparation, the weld is referred to as a square weld since the edges are square and flush, not beveled. If both the top and bottom sides of the two metal pieces are to be welded, then it is referred to as a double square weld. In addition, the term "groove" may also be included to identify welds where the edges of both pieces are butted up to each other, as illustrated in Figure 12.2. This is not the case in plug or slot weld. In addition, a *backing weld* refers to a weld placed on the inside of a pipe to prevent the shielding gases from escaping when welding the front side of the weld. Moreover, another type of weld is referred to as a *back weld* which is placed on the opposite side of the main weld after it has been welded.

After reviewing Table 12.1, it can be seen that there are bevel joints and V-joints as well as J-joints and U-joints which would seem to be the same. However, it depends upon the edge that is prepared for welding. For instance, if one edge of the two pieces of metal to be joined is ground at an angle, then it is referred to as a single bevel. If the bottom of that same piece of metal is ground at an angle as well, then it is referred to as a double-bevel joint. On the other hand, if the top edges of both pieces of metal being welded together are ground at an angle, then it forms a v shape, so it is called a Single-V joint. If the bottom side edges are ground at an angle as were the tops sides of both metal pieces, then it is referred to as a Double-V joint.

TABLE 12.1 Weld type recommendations for material thickness.

Weld Type	Thickness
Square butt joint	Up to ¼ in (6.35 mm)
Single-bevel joint	3/16–⅜ in (4.76–9.53 mm)
Double-bevel joint	Over ⅜ in (9.53 mm)
Single-V butt joint	Up to ¾ in (19.05 mm)
Double-V butt joint	Over ¾ in (19.05 mm)
Single-J joint	½–¾ in (12.70–19.05 mm)
Double-J joint	Over ¾ in (19.05 mm)
Single-U joint	Up to ¾ in (19.05 mm)
Double-U joint	Over ¾ in (19.05 mm)
Flange (edge of corner)	Sheet metals less than 12 gauge (0.1046 in or 2.657 mm)
Flare groove	All thickness

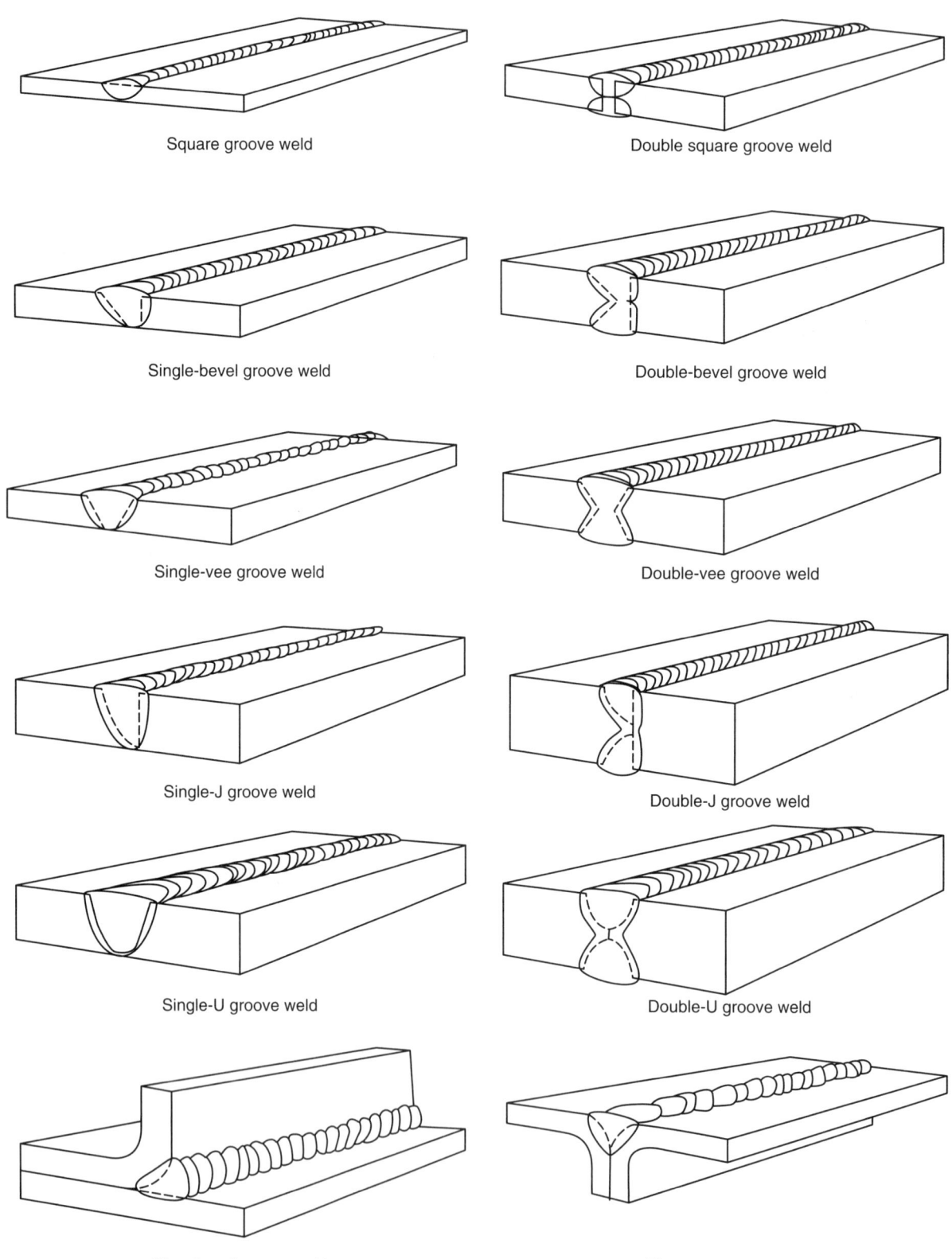

Square groove weld

Double square groove weld

Single-bevel groove weld

Double-bevel groove weld

Single-vee groove weld

Double-vee groove weld

Single-J groove weld

Double-J groove weld

Single-U groove weld

Double-U groove weld

Flare-bevel groove weld

Flare-vee groove weld

FIGURE 12.2   The main types of welds used to fabricate metal pieces.

This same type of thinking holds true for the J- and U-joints. If the top edge of only one piece of metal is ground down in the shape of a J then it is called a Single-J joint. When the top and bottom of only one piece of metal is ground down in the shape of the letter J, then it is called a Double-J joint. Now, if the top edges of both pieces of metal are ground down in the shape of a J, they create a U-shape and are thus called a Single-U joint. If both the top and bottom edges of both pieces of metal are ground down in the shape of a J, then they form a Double-U joint. Refer to Figure 12.2 for further clarification of the various types of welds.

Surface preparation for bevel, V, U, and J-joints can be done in a number of ways. Depending upon the tools that are available and the requirements for the job, bevels can be cut with an oxyacetylene torch, plasma cutter, air carbon arc cutter/gouger, handheld angle grinder, turned on a lathe, and a host of specific pipe and plate beveling machines or fixtures that hold the various cutting devices at adjustable angles to obtain the correct setting. Refer to Figures 12.3 to 12.5.

SYMBOLS FOR WELDS

On the blueprints that a pipefitter will be given are major symbols that consist of lines, arrows, circles, numbers, and letters for each weld that needs to be made; refer to Figure 12.6. Initially, this looks quite confusing and can be overwhelming so this chapter will further break down what everything means.

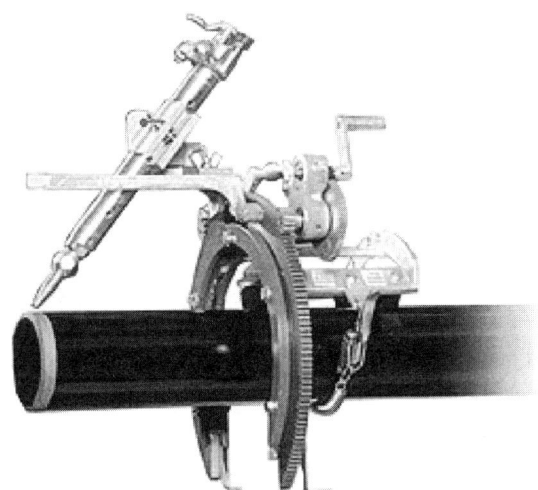

FIGURE 12.3 Pipe beveling attachment for oxyacetylene cutting (Red-D-Arc).

FIGURE 12.4 Portable edge beveling tool (Metabo).

FIGURE 12.5 Automatic pipe beveling and cutting machine used in the field on a pipeline (E.H. Wachs).

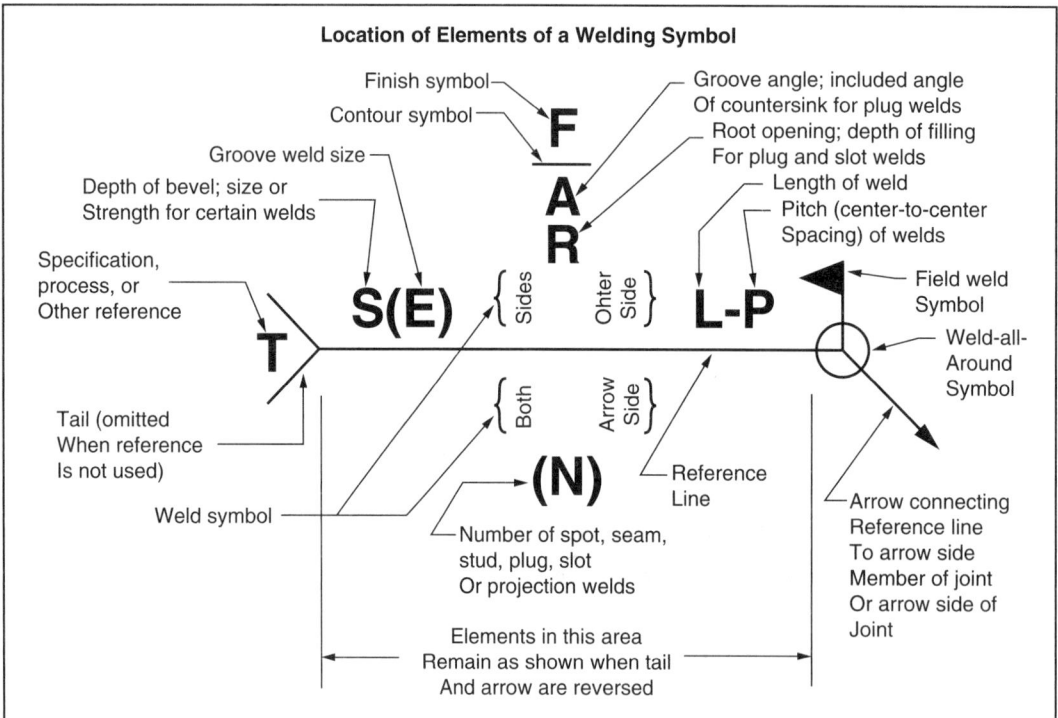

FIGURE 12.6 Elements that comprise a welding symbol.

The welding symbol shown in Figure 12.6 will be further subdivided from left to right. The far left T symbol represents the type of welding process that would be used. These processes are listed in Table 12.2.

The S depicts how deep a bevel should be cut into the metal piece or pieces. Typically, the bevel is not beveled throughout the entire thickness of the metal pieces. The part of the edge that is not ground is termed the root face or land. Not only is it safer to have a blunt edge to butt together (no sharp metal edge), but there needs to be some thickness of base metal for the weld to attach to.

The (E) in Figure 12.6 is the T plus the welding penetration. Therefore, if the bevel is ¾-inch deep on a 1-inch thick piece of metal plate and the penetration of the weld is 1/8 inch, then the (E) would be ¾" + 1/8" = 7/8". If no weld size is shown the weld should be the complete joint penetration or 1-inch thick.

TABLE 12.2 Welding process letter designations.

Major Classification	Description of Process	Letter Designation
Brazing and gas welding	Torch brazing	TB
Brazing and gas welding	Oxyacetylene welding	OAW
Arc welding	Shielded metal arc welding or stick	SMAW
Arc welding	Gas metal arc welding or MIG	GMAW
Arc welding	Gas tungsten arc welding or TIG	GTAW
Arc welding	Flux core arc welding	FCAW
Arc welding	Plasma arc welding	PAW
Arc welding	Submerged arc welding	SAW
Arc welding	Stud welding	SW
Arc cutting	Air carbon arc cutting	AC
Arc cutting	Plasma arc cutting	PAC
Oxygen cutting	Oxyacetylene cutting	OAC
Oxygen cutting	Metal powder cutting	POC

At the top part of the welding symbol, there is an F for the finish. This can be labeled in two different ways. There can be a straight line, which would mean to finish it flat or a line sunken in for concave or a line bowing out for convex. A letter can also be used to determine how it is finished. The letter G would designate a ground finish, the letter M would be a machine finish, and the letter C would mean that the surface of the weld was cleaned up with a chipping hammer.

The letter A below the finish mark would signify the angle of any type of groove that is cut into the edges of the weld. This would be the included angle, for example, 30 degrees for a one-sided bevel and 60 degrees for a V weld where the edge on both metal pieces that are butting together and ground at a 30-degree angle, thereby adding up to 60 degrees.

Underneath the A is the R which would represent the root opening or the space the two pieces to be joined are from each other. Therefore, if the pieces are butted tightly together then that would be labeled zero; however, if there was a gap then that space would be written there, such as 1/16 inch or 1/8 inch.

The letter L would represent the length of the weld if it did not go the entire length of the pieces being joined. If it did go the entire length, then this would not be labeled.

As for the letter P, it would also not be labeled if the weld went the entire length of the pieces being welded. However, if the weld went a few inches, stopped, and then started again for a few more inches (stitch weld), then the spacing between the centers of each weld would be listed for the letter P.

If there is a circle where the arrow line intersects with the reference line (the horizontal line of the welding symbol) then the two pieces will be welded all the way around on all four sides.

When a flag appears on the welding symbol, then this weld will be completed onsite where the piping or part will be installed. If not, then it can be welded in a factory.

The (N) letter just signifies how many spot welds would be made along the length of the two pieces being welded together. It would also list the spacing between the spot welds. This would only be listed if the pieces were spot welded together.

One of the most important things to consider is where the arrow is pointing to on the pieces to be welded. For instance, if the arrow is pointing to where a fillet weld will be made on a tee-joint, then a fillet symbol would be drawn on the bottom side (which is the arrow side because the arrow is pointing down). If the other side of the tee-joint is to also be fillet welded, then a fillet symbol would be drawn on top of the horizontal reference line as well. This concept is illustrated in Figure 12.7 with a tee-joint and a lap joint. The triangle symbol on the welding symbol represents a fillet weld.

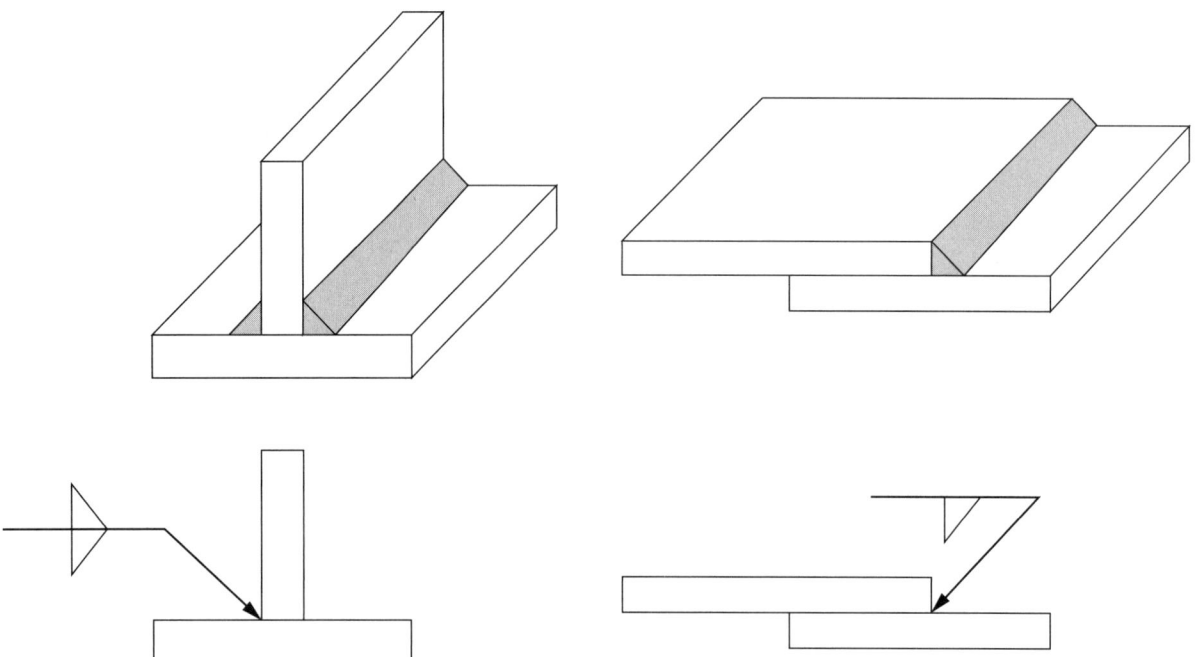

FIGURE 12.7   Representations of welding symbols.

Representations of weld types and supplementary symbols are listed in Figure 12.8.

More examples of how to read a welding symbol are listed in Figure 12.9. The numbers 2 to 5 listed in Figure 12.10 represent the length of each fillet weld (2 inches) on the tee-joint and that they are spaced 5 inches on center from each other.

Type of Weld								Supplementary Symbols			
Back or Backing	Fillet	Plug or Shot	Groove					Weld All Around	Field Weld	Contour	
			Square	V	Bevel	U	J			Flush	Convex
⌒	◺	▭	‖	∨	⋁	⋃	∪	⊘—	⌐	—	⌢

FIGURE 12.8   Symbols for types of welds and supplementary symbols.

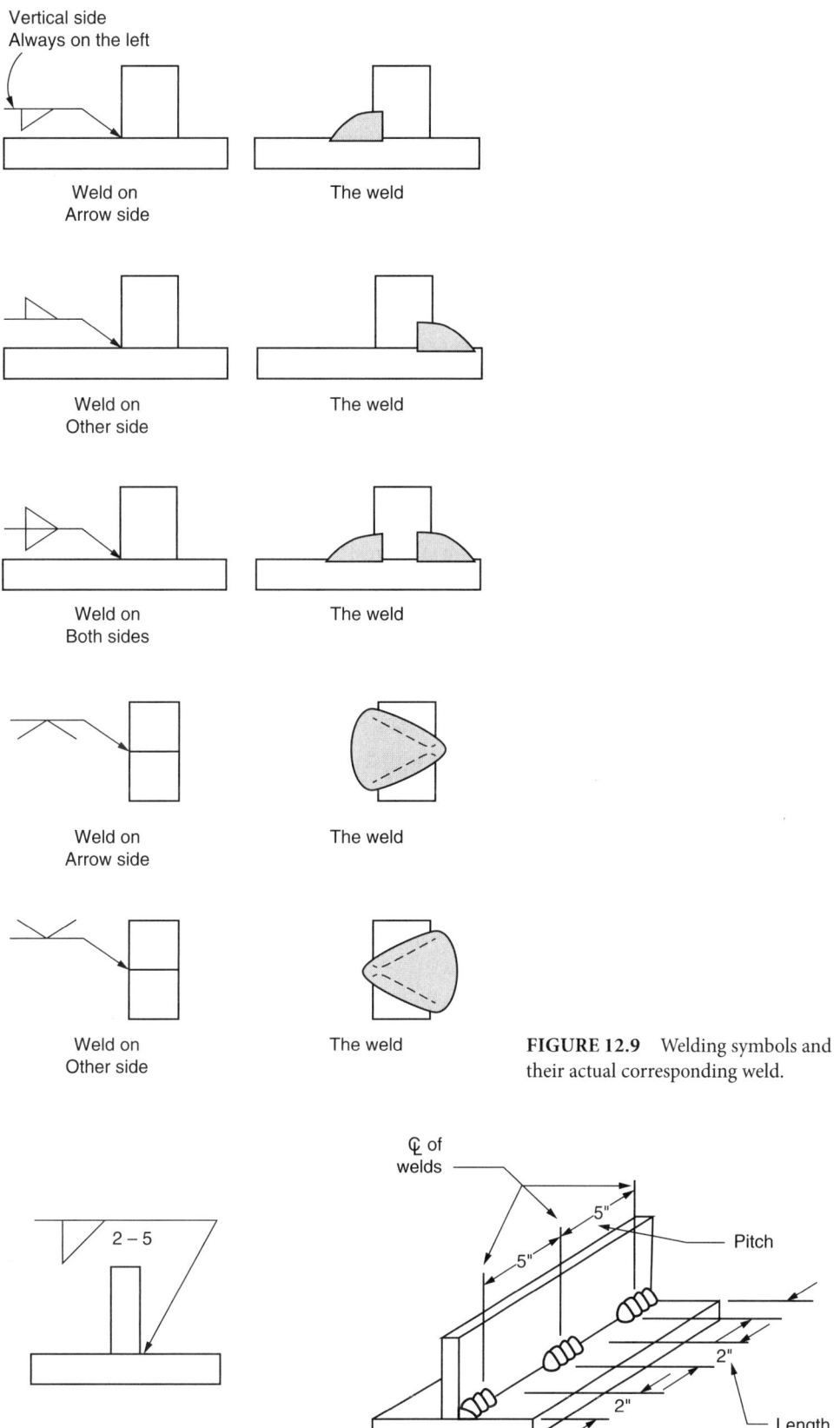

FIGURE 12.9 Welding symbols and their actual corresponding weld.

FIGURE 12.10 2-inch long fillet welds spaced 5 inches on center from each other.

REVIEW QUESTIONS

1. Which type of welding joint is typically used to weld pipe together?

    a. tee

    c. lap

    e. edge

    b. butt

    d. corner

2. Which of the following is considered the weakest of the welding joints?

    a. tee

    c. lap

    e. edge

    b. butt

    d. corner

3. Which welding joint would be considered the strongest?

    a. tee

    c. lap

    e. edge

    b. butt

    d. corner

4. When this joint is put together it forms the shape of a capital letter and would be comprised of fillet welds.

    a. tee

    c. lap

    e. edge

    b. butt

    d. corner

5. Which type of welding joint is the simplest to make?

    a. tee

    c. lap

    e. edge

    b. butt

    d. corner

6. Which type of weld is only good for materials up to ¼-inch thick?

    a. Double-U

    c. Single bevel

    b. Single-J

    d. Square butt

7. Which type of weld is best for welding materials over ¾-inch thick?

    a. Double-U

    c. Single bevel

    b. Single-J

    d. Square butt

8. What is the name of the weld made inside a pipe after the main weld is completed?

   a. plug
   b. backing
   c. back
   d. none of these

9 What is the name of the weld where one piece of the metal is ground at an angle on the top and the bottom of the edge?

   a. Single-V
   b. Double-V
   c. Single bevel
   d. Double bevel

10. What is the name of the weld where the top edges of both pieces of metal being welded together are ground in the shape of the letter J?

   a. Single-J
   b. Double-J
   c. Single-U
   d. Double-U

11. Which of the following is NOT a recommended tool for preparing a Single-V weld?

   a. oxy-acetylene torch
   b. air carbon arc cutter
   c. grinder
   d. lathe
   e. sanding block

12. This picture illustrates what type of welding joint?

   a. tee
   b. butt
   c. lap
   d. corner
   e. edge

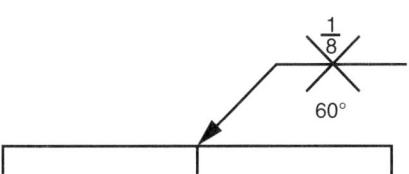

13. According to the weld symbol in the illustration, where would the weld be made?

   a. top surface
   b. bottom surface
   c. edge
   d. both the top and bottom surfaces

14. What does the 1/8 inch represent in the illustration?

   a. width of the weld
   b. depth of the weld
   c. root spacing
   d. root face length

15. What type of weld is represented in the illustration?

    a. Single bevel   b. Double bevel

    c. Double-U   d. Double-V

16. The (E) on the welding symbol diagram represents:

    a. depth of bevel   b. groove angle

    c. number of spot welds   d. groove weld size

17. The S on the welding symbol diagram represents:

    a. depth of bevel   b. groove angle

    c. number of spot welds   d. groove weld size

18. The N on the welding symbol diagram represents:

    a. depth of bevel   b. groove angle

    c. number of spot welds   d. groove weld size

19. What would the letter G represent on a welding symbol?

    a. groove weld   b. gouge joint preparation

    c. grind to specified angle   d. ground finish

20. What does the letter P represent on the welding symbol diagram?

    a. welding process

    b. part thickness

    c. center-to-center spacing of the stitch welds

    d. none of these

ANSWERS TO REVIEW QUESTIONS

1. b	2. d	3. c	4. a	5. b
6. d	7. a	8. c	9. d	10. c
11. e	12. b	13. d	14. c	15. d
16. d	17. a	18. c	19. d	20. c

Chapter 13 ———————————————○
READING PIPING PRINTS

Performance Objectives

After studying this chapter you will (be able to):

1. Know the difference between the various piping schematics.

2. Understand the importance of orthographic projections.

3. Explain the importance of a piping spool.

4. Describe the importance of a bill of materials.

5. Identify the various symbols used on piping drawings.

PIPING PRINTS AND DIAGRAMS

In order for pipefitters to properly install pipes, valves, fittings, etc., correctly they must be able to read the plans that have been drawn up to identify these materials' proper type, length, location, and placement. Furthermore, pipefitters will need to know if insulation would be wrapped around the pipes, if electronic devices would be installed to monitor the control of the flow as well as a host of other things. The common plans that pipefitters would be following are schematic diagrams, flow diagrams, piping and instrumentation diagrams (P&ID), isometric drawings, elevations, spool drawings, and bill of materials. All these drawings, or blueprint copies used in the field, provide pipefitters with enough information to make a piping system that was originally designed by licensed or certified engineers.

SCHEMATIC DIAGRAM

This is an initial diagram of the layout and operation of a system. It includes the various components that would be required to make the system function. Specific lengths, placement, etc., are not included on these diagrams. Schematic diagrams are primarily used for the initial planning of a system as illustrated in Figure 13.1.

FLOW DIAGRAM

This type of diagram provides more information than a basic schematic diagram. Although nothing is drawn to scale, it does have arrows to show the flow of the fluids through the system and symbols that are used in P&ID drawings are being introduced. This diagram is just an overall picture of all the components that are required for the system plus additional valves and other devices that are needed to run the system in the desired temperature and pressure ranges. However, none of the specific requirements for these devices is listed in this diagram as noted in Figure 13.2.

PIPING AND INSTRUMENTATION DIAGRAM

A P&ID contains all the standardized symbols used in the industry and has greater detail than the previous diagrams. As illustrated in Figure 13.3, the actual pipe type is listed as 4" Sch. 40 CS (which is 4 inches in diameter, Schedule 40, carbon steel pipe), a legend is given with utility connections, instrumentation balloons are

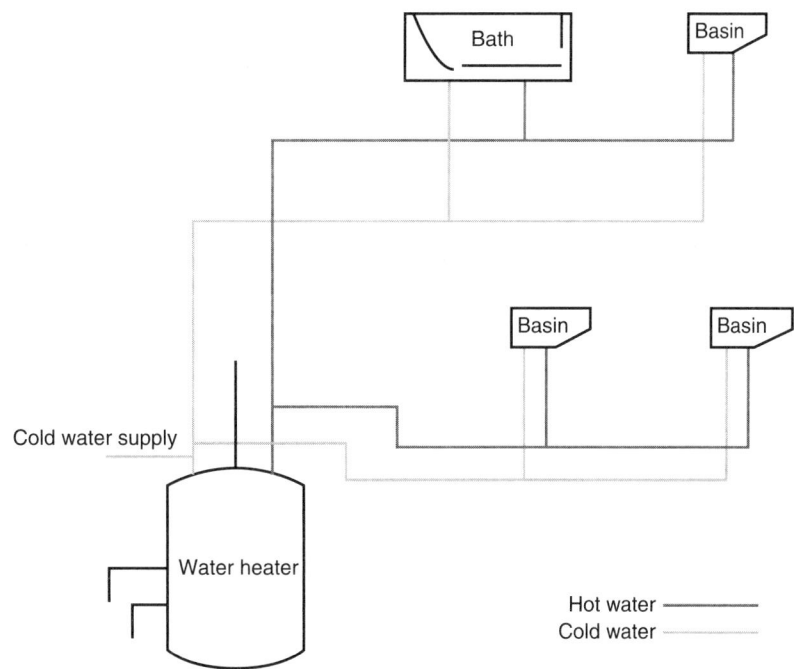

FIGURE 13.1 Basic schematic diagram.

Process Flow Diagram or PFD

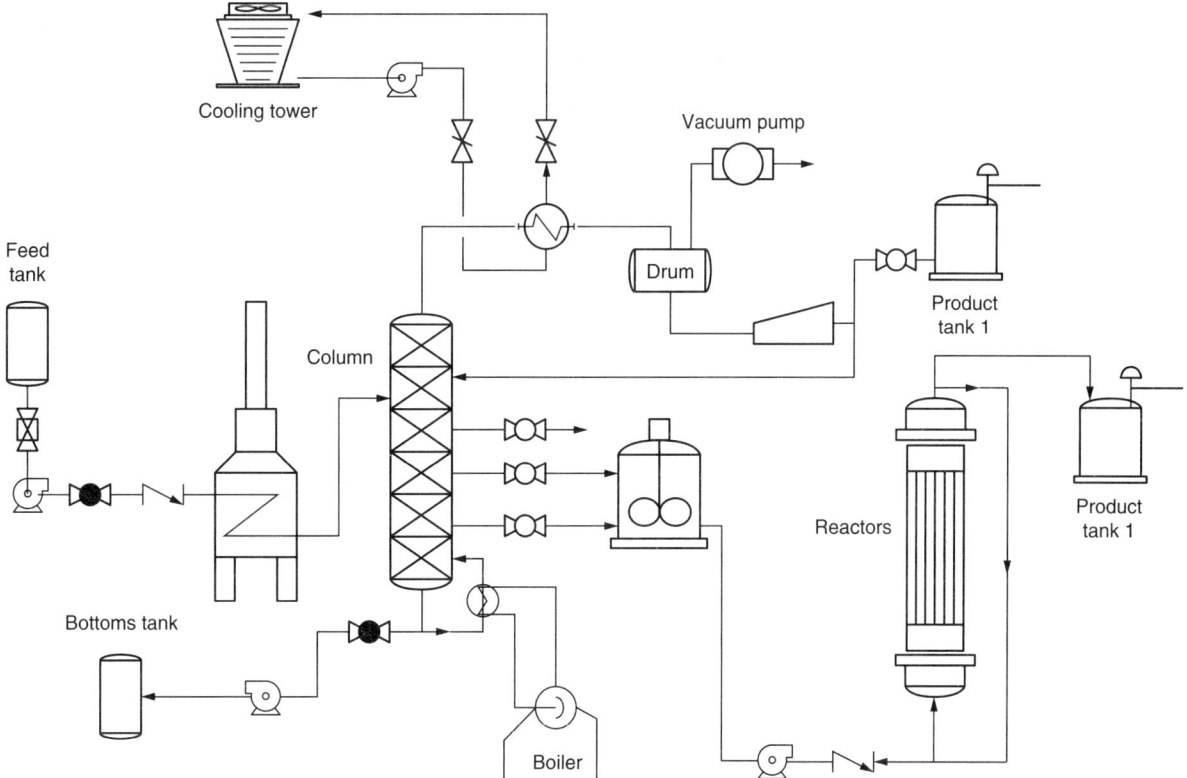

FIGURE 13.2 Flow diagram example.

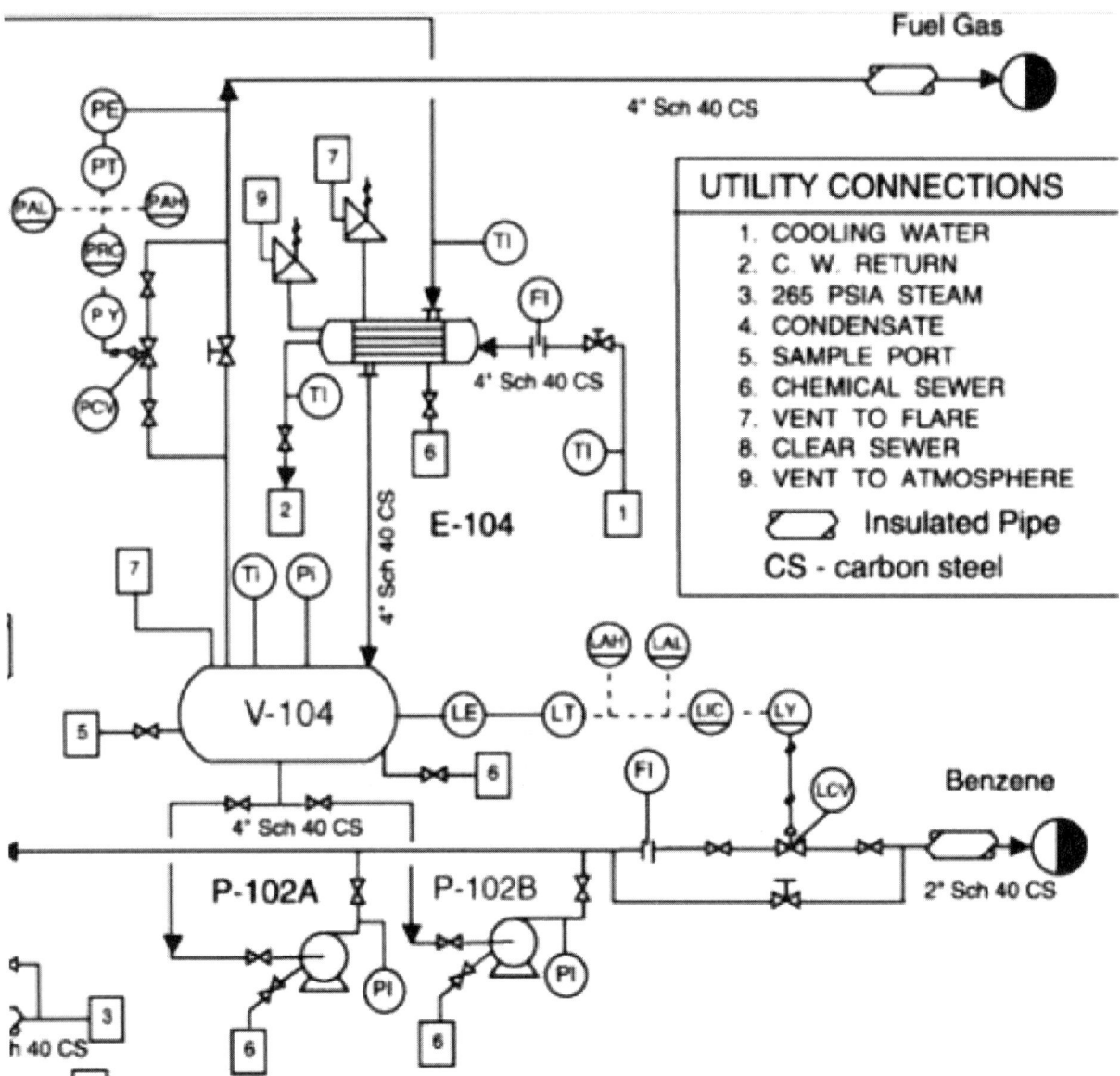

FIGURE 13.3 An example of a pipe and instrumentation diagram (P&ID).

listed (letters and/or numbers with a circle around them), solid and dashed lines (solid lines represent pipes, dashed are electrical lines), and so forth. Again, this is not drawn to the actual scale of the system nor does it list the actual lengths of the pipes to various components as shown in Figure 13.3.

ISOMETRIC DRAWING

An isometric drawing is a way of making a drawing look three-dimensional for better clarification as the naked eye would see it. A person would be viewing an object or piping system as if they were looking at it from a corner where two sides could be viewed as well as part of the top. In order to achieve this perspective,

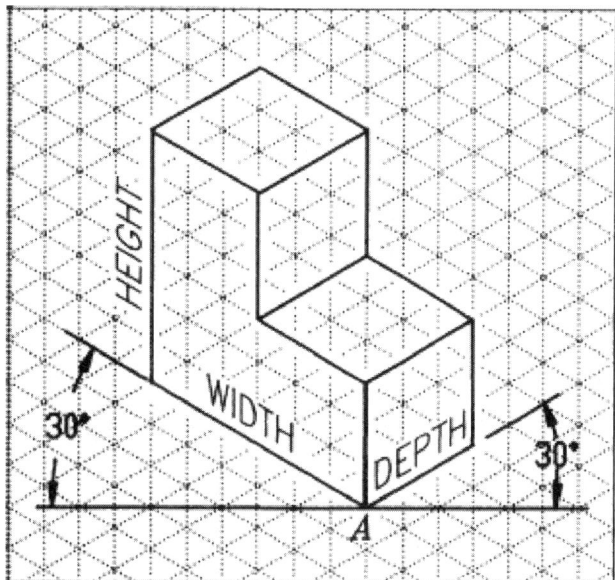

FIGURE 13.4   An example of an isometric drawing.

all the horizontal lines are drawn back at a 30-degree angle while the vertical lines do not change as shown in Figure 13.4.

This is an excellent way to illustrate piping systems and all their components, especially when they span several levels as in a multistory building or refinery. An example of an isometric piping drawing is shown in Figure 13.5.

ORTHOGRAPHIC PROJECTION

In orthographic projection an object or system would be drawn as if a person was looking at it straight from the front, right side, top, left side, bottom, and back. In most cases, three views of an object are enough to represent the particular characteristics of it so the front, top, and right side are typically drawn. Moreover, the front view is not necessarily the front of an object. The front view is the view that would represent the angle of an object, so it could be recognized. For instance, if a person looks at the front of the car, they would see the grill, some of the hood, the windshield, and the tires would be drawn two dimensionally as little rectangles underneath the car. However, if the side of a car was drawn, a person would see two round tires and wheels, the doors, windows, and the outline of the car. In that case, the side of the car would be the front view since it would visually represent a car better than any of the other views as illustrated in Figure 13.6.

SECTIONALS AND ELEVATIONS

Orthographic projection is important with piping drawings because it can add an extra view to a complicated assembly. This view can be a cutaway or an enlargement of a section of a piping system drawn orthographically, so it can be seen at a better angle to show more detail in its true size and shape. These drawings are called sectionals and elevations, respectively.

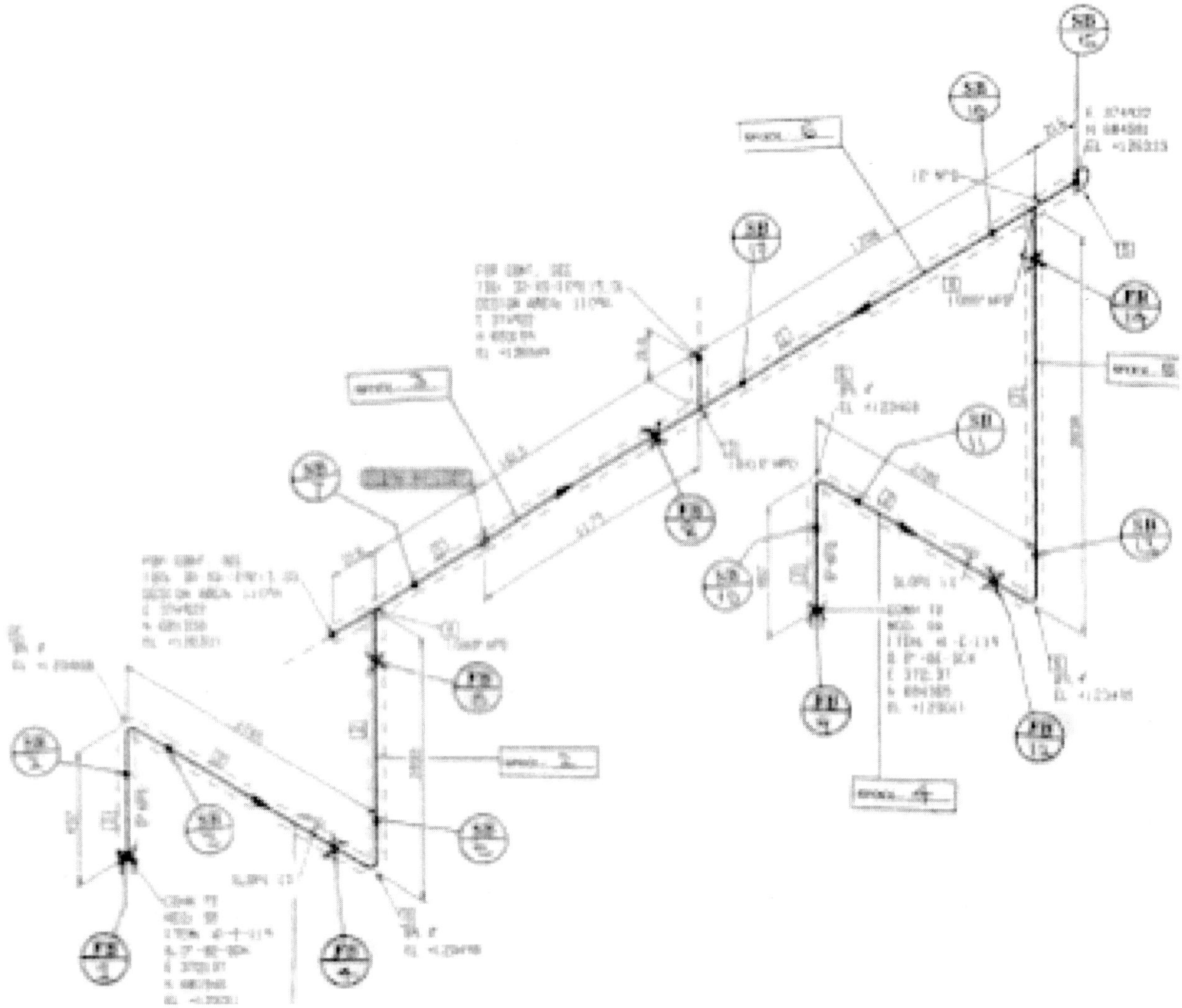

FIGURE 13.5 Isometric drawing of a piping system.

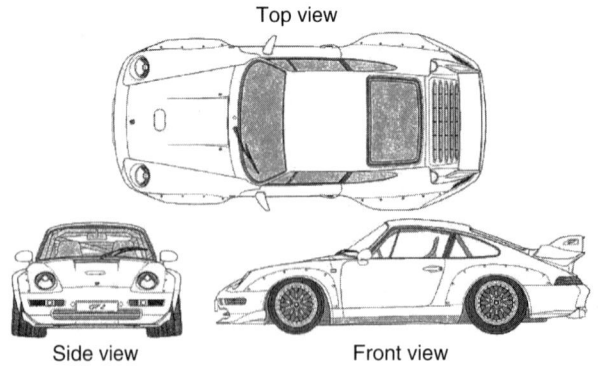

Top view

Side view   Front view

FIGURE 13.6 Orthographic drawing of a car.

PIPE SPOOLS

It is much easier and faster to make sections of a piping system in a fabrication facility and then have it trucked out to the site to be assembled. A fabrication facility typically has jigs and fixtures for exact setups and automated welders to improve quality as well as a better, more controlled working environment. These prefabricated sections of a piping system are called spools and are numbered on the piping plans. Plans for each spool are required so the fabrication facility can build it to specifications; refer to Figure 13.7.

BILL OF MATERIALS

The bill of materials is a list of all the components that will be assembled to make a piping system or whatever is being manufactured according to the plans. The bill of materials will list the quantity of these components, specifications, and item number. Each drawing spool will have its own bill of materials listed on the specific detail plan. In addition, the master set of plans, typically named an assembly drawing, will have a list of all the components required to build the entire system. An example of a bill of materials is shown in Figure 13.8. A list of piping symbols are listed in Tables 13.1 to 13.4.

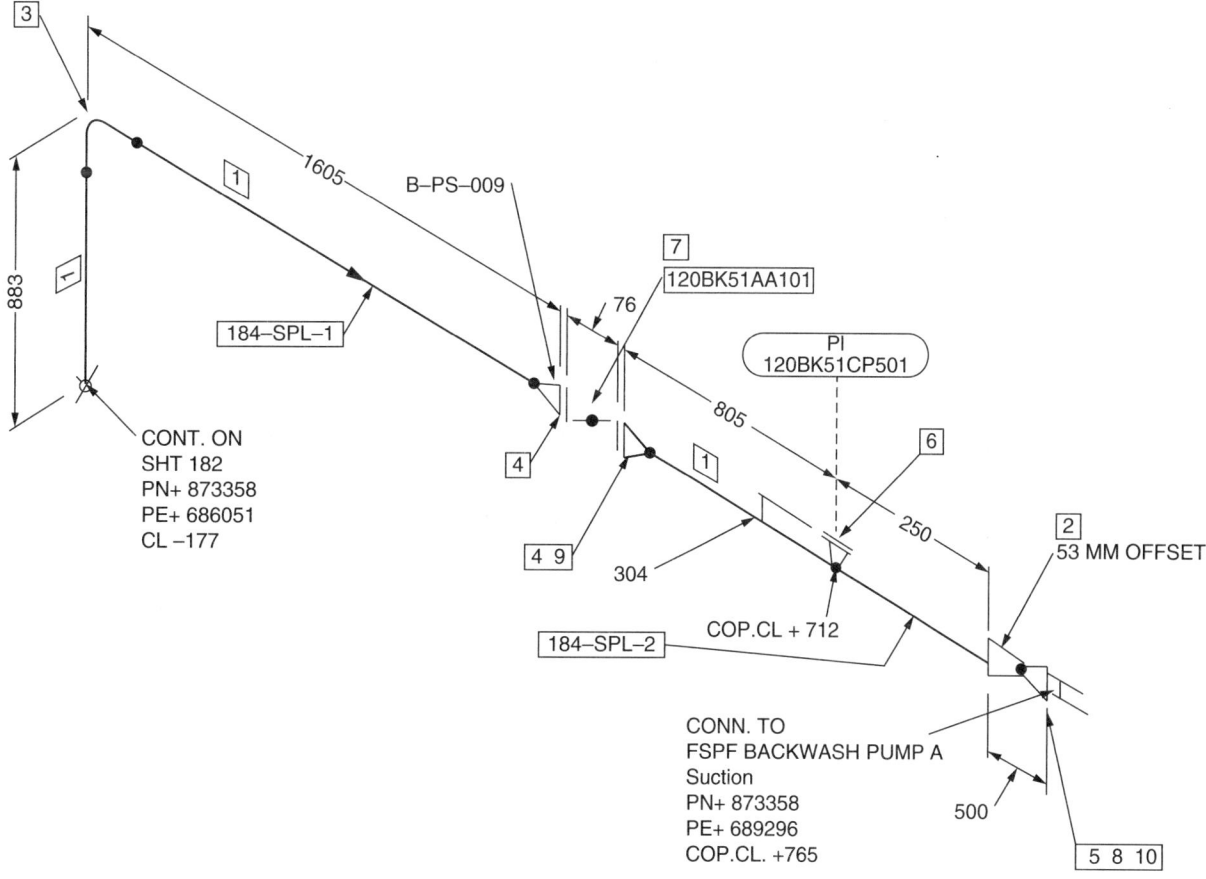

FIGURE 13.7 Pipe spool example.

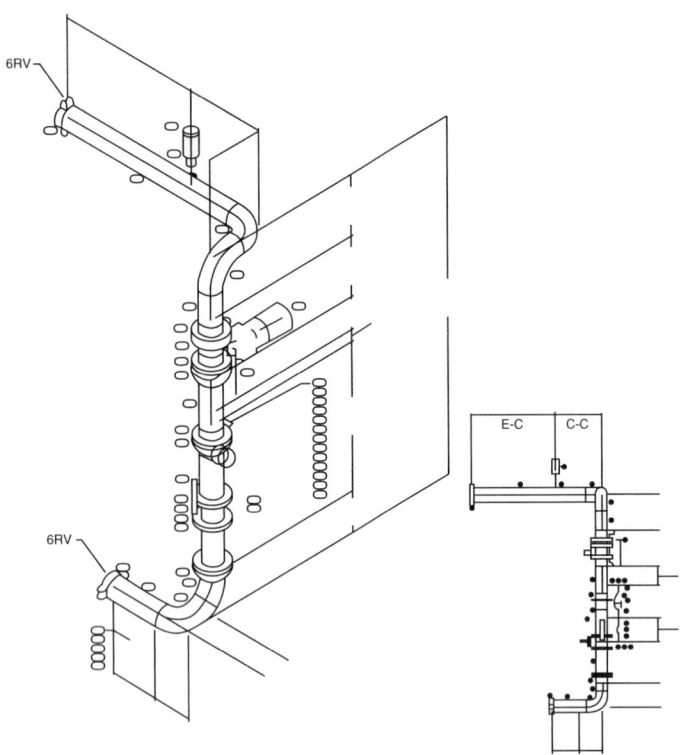

BILL OF MATERIALS			
Mark	Qty	Size	Description
1	1		
2	1		
3	1		
4	1		
5	1		
6	1		
7	1		
8	1		
9	1		
10	1		
11	1		
12	1		
13	1		
14	1		
15	1		
16	1		
17	1		
18	1		
19	1		
20	1		
21	1		
22	1		
23	1		
24	1		
25	1		
26	1		
27	1		
28	1		
29	1		

FIGURE 13.8   Bill of materials for a piping system spool.

TABLE 13.1   Piping valve symbols.

In-line valve	Check valve	Diaphragm valve	Plug valve	Statically loaded	Quick opening
3-way valve	Check valve	Diaphragm valve	Plug valve	Spring loaded	Quick opening
4-way valve	Screwdown valve	Diaphragm valve	Plug valve, straight through	Spring loaded	Quick closing
Screw-down valve	Float operated valve	Wedge gate valve	3-way plug valve	Remote control	Quick closing
Lock-shield valve	Float operated valve	Parallel side valve	Plug valve, T-port	Diaphragm	Connecting unit
Reel valve	Flanged valve	Gate valve	Plug valve, T-port	Diaphragm, positioner	Connecting unit
Relief valve	Flanged valve	Ball valve	3-way plug valve	Chain operated	Connecting unit
Relief valve	Butterfly valve	Ball valve	3-way plug valve	Gear operated	Motor element
Relief valve	Butterfly valve	Ball valve	3-way plug valve	Solenoid	Motor element, opens on failure
Relief valve	Globe valve	Powered control valve	3-way plug valve, T-port	Weight loaded	Motor element, closes on failure
Angle valve	Globe valve	Powered control valve	3-way plug valve, L-port	Weight loaded	Motor element, retains position on fail
Angle valve	Globe valve	Powered control valve	Mixing valve	Weight loaded	Motor element, safe direction
Angle valve	Needle valve	Relief angle valve, pressure	Characterized port valve	Float operated	Regulating
Angle valve	Needle valve	Relief angle valve, vacuum	Manual isolation	Float operated	
Angle valve	Needle valve	Reducing valve	Power signal	Dash-pot	
	Needle valve	Reducing valve		Dash-pot	
				Piston	

TABLE 13.2   Piping symbols for various fittings (Werner Sölken).

Image	Fittings	Butt weld symbol	Socket weld symbol	Threaded symbol	Fittings	Image
	Elbow 90°				Elbow 90°	
	Elbow 45°				Elbow 45°	
	Tee equal				Tee equal	
	Tee reducing				Tee reducing	
	Cap				Cap	
	Reducer concentric		...	...	Reducer concentric	...
	Reducer eccentic		...	...	Reducer eccentic	...
Image	Fittings	Butt weld symbol	Socket weld symbol	Threaded symbol	Fittings	Image

TABLE 13.3   Piping symbols for various flanges (Werner Sölken).

Flanges	Weld neck	Socket weld	Threaded	Slip-On	Lap-joint	Blind	Flanges
Symbol							Symbol
Image							Image
Flanges	Weld neck	Socket weld	Threaded	Slip-On	Lap-joint	Blind	Flanges

TABLE 13.4 Piping symbols for various valves (Werner Sölken).

Image	Valves	Butt weld symbol	Flanged symbol	Socket or threaded symbol	Valves	Image
	Gate				Gate	
	Globe				Globe	
	Ball				Ball	
	Plug				Plug	
	Butterfly			...	Butterfly	
	Needle				Needle	
	Diaph	...			Diaph	
	Y-type				Y-type	
	Three Way				Three Way	
	Check				Check	
	Bottom	...		...	Bottom	
	Relief	...		...	Relief	
	Control straight	...		...	Control Straight	
	Control angle	...		...	Control Angle	
Image	Valves	Butt weld symbol	Flanged symbol	Socket or threaded symbol	Valves	Image

REVIEW QUESTIONS

1. Which of the following diagrams is just an initial layout and operation of a system?

   a. flow

   b. schematic

   c. P&ID

   d. isometric

2. This type of diagram has the most detail and would even list the type of pipe, location of electrical lines, but is not drawn to scale.

   a. flow

   b. schematic

   c. P&ID

   d. isometric

3. This type of diagram has arrows on it, components, but not specific requirements for each.

   a. flow

   b. schematic

   c. P&ID

   d. isometric

4. What does a bow tie symbol represent on a P&ID?

   a. fitting

   b. flange

   c. valve

   d. pipe

   e. vessel

   f. pump

5. Which type of valve has a dark circle in the middle of it?

   a. gate

   b. globe

   c. check

   d. ball

6. What angle are horizontal lines made on an isometric drawing?

   a. 15

   b. 30

   c. 45

   d. 60

   e. 90

7. Orthographic projections typically represent which sides of an object?

   a. front, top, right side

   b. front, top, bottom

   c. top, left side, front

   d. top, front, isometric

8. The front view of an orthographic projection is_____.

   a. the interior detail of an object

   b. the overhead view of an object

   c. functioning part of an object

   d. the side that best identifies it

9. Which type of drawing is a sectional?

   a. orthojectional projection   b. isometric

   c. a cutaway   d. an enlargement

10. What is the name for a prefabricated section of a piping system?

    a. bill of materials   b. spool

    c. sectional   d. elevation

ANSWERS TO REVIEW QUESTIONS

1. b	2. c	3. a	4. c	5. b
6. b	7. a	8. d	9. c	10. b

Chapter 14
OXYACETYLENE WELDING

Performance Objectives

After studying this chapter you will (be able to):

1. List the various personal protective equipment used when welding.

2. Understand the safety precautions involved with welding.

3. Explain the advantages and disadvantages of oxyacetylene welding.

4. Describe how to set up an oxyacetylene rig for welding.

5. Know the difference between oxyacetylene filler rods.

6. Demonstrate how to obtain the correct welding flame with an oxyacetylene rig.

7. Describe the differences between the three types of flames.

8. Troubleshoot problems that may arise with oxyacetylene welding.

There are several different processes used to weld pipe, and depending on the material, specifications, location, and other requirements a pipefitter should be able to weld in a variety of ways. Before reviewing the various welding methods, a quick overview on safety will be presented.

WELDING SAFETY

Most welding processes produce heat, smoke, fumes, sparks, and other potential threats that may be hazardous to a person's health. Therefore, proper personal protective equipment (PPE) should be worn to protect the welder from many of these hazards. Some forms of PPE are as follows:

1. Welding leathers are thick and will prevent welding sparks from burning through clothing and skin. Special welding jackets, aprons, sleeves, gloves, capes (half jackets), and leggings are available in leather as well as in other fire-resistant materials.

2. Thick cotton denim shirts and jeans are best to wear since clothing with a percentage of plastic in it tends to melt. Shirts should not have open pockets that welding sparks and slag could fall into and eventually burn through to the skin. Open cuffs should not be worn as well.

3. Steel-toed boots should be worn to prevent hot, heavy pipe from injuring a foot when falling.

4. A welding cap should be worn to protect the top of the head from burning sparks, etc. Sometimes do-rags, skull caps, beanies, etc., are also worn so they do not interfere with the welding helmet.

5. A welding helmet should be worn when using most of the welding processes, but in some cases goggles with tinted lenses will suffice when brazing or welding with an oxyacetylene gas rig. However, it is recommended that a clear plastic face shield be worn over the goggles for maximum safety. Welding helmets are designed to allow for a shaded piece of glass and its front and back protecting clear cover lenses to be replaced with ease. Lens shades will depend on how heavy or intense the welding arc is and are numbered. A shade numbered 4, 5, or 6 would work for oxyacetylene gas welding, while darker shades (9–12) would be required for arc welding. The higher the number, the darker the lens and more protection against light.

To select the correct shade, weld with the recommended number lens. If light spots are seen after welding, step up to a darker shade; if dark spots are seen after welding, go down to a lighter shade.

Safety Precautions

It should be noted that no one should weld or cut on tanks, vessels, or other containers that may have stored hazardous substances and fumes that could either explode with a spark or cause toxic fumes to be released. A well-ventilated area is always required when welding. Welding in a tank with an argon shielding gas could eventually remove all the oxygen in the tank for the welder to breathe properly. In addition, welding machines require a lot of current so all the cables should be checked for cracks and cuts and replaced when necessary. A dry environment is also needed for welding so a welder does not get shocked. Do not stand in water when welding or wear damp gloves.

Oxygen and Acetylene Safety

All welding cylinders (also referred to as bottles or tanks) should be labeled with the correct gas that is inside them. Acetylene is the combustible fuel gas and must be treated with care. If an acetylene cylinder is laid on its side then the acetone inside will become unstable and could explode. The acetylene tanks are easier to identify because they are typically shorter and wider. They are typically wider because they are filled with some type of monolithic porous honeycomb filler that allows the acetylene gas to be more stable when mixed with acetone. Typically, acetylene cylinders are red, but older ones came in gray and black. Color coding of cylinders is not a good way to identify gases because there is no standard in the United States. Always check the label that is wrapped around the top of the cylinder for proper identification. Do not use the cylinder if the label is illegible.

Oxygen is a colorless, odorless, and tasteless gas that is not flammable. However, it allows fuel gases to burn better than in air. When mixed properly with acetylene, oxyacetylene can easily weld and cut steel as well as a variety of other metals. Remember to keep oil and grease away from pressurized oxygen or it will explode. This means not to use oxygen like compressed air since it will increase the volatility of sparks, etc., and lead to fires and explosions.

Gas cylinders should always be stored upright and moved in the same manner. They should be secured with chains or heavy duty straps to the wall or cart that they are stored in. Always screw on the removable metal cap when not using or moving the gas cylinder. In this way, if the cylinder falls it will not break off the valve which could cause the pressurized gas cylinder to shoot into the air like a rocket. The pressure is high enough to cause the heavy metal cylinder to go through concrete walls. Extreme caution must be maintained when working with gas cylinders. Once gas cylinders are set up with hoses and regulators, soapy water should be sponged around the fitting to check for leaks. A leaky fuel gas fitting could fill up a room over-night and a spark from a light switch the next morning could ignite it causing an explosion just like a natural gas leak in a house. Always check for the correct working pressure for various gases. Acetylene can only be pressurized to a 15-psi working pressure and should not exceed that because the acetone will also be released from the cylinder deteriorating weld quality and making its gas less stable where it could explode. Notice the difference in size and shape of an acetylene cylinder verses an oxygen cylinder as show in Figure 14.1.

The first welding process used commercially was oxyacetylene. It is good for many applications, in particular:

- Cast iron
- Wrought iron
- Low-alloy steels
- Copper
- Bronze

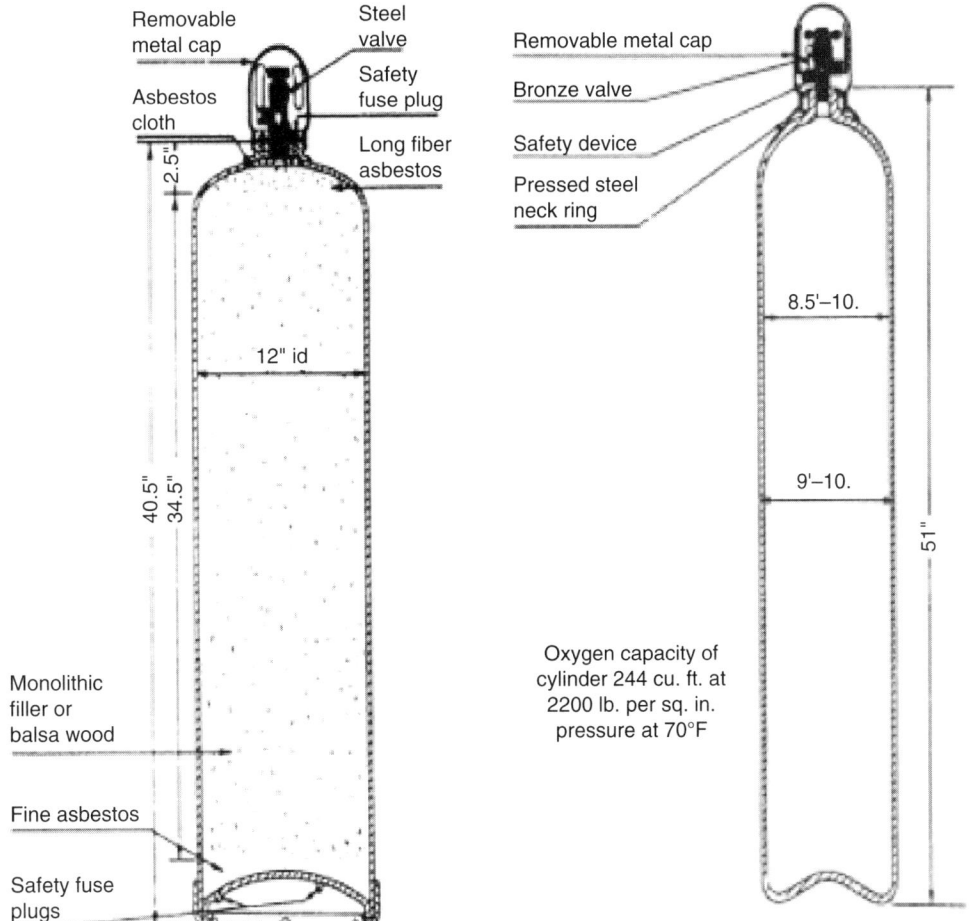

FIGURE 14.1 A cutaway view of acetylene and oxygen cylinders respectively.

Except for repair and maintenance work, oxyacetylene welding has been replaced by various arc-welding processes, including shielded-metal-arc welding (or stick welding), gas-metal-arc welding (or MIG welding), and tungsten-metal-arc welding (or TIG welding). Although relegated to a minor role in welding, oxyacetylene is still used for a wide variety of other non-welding uses, such as:

- Cutting
- Preheating
- Postheating
- Flame hardening
- Case hardening
- Braze welding
- Brazing
- Soldering
- Descaling

Advantages of oxyacetylene welding (OAW) include the following:

- Self-contained and easily portable equipment
- Widely available equipment
- Relatively inexpensive equipment
- Easy to learn

Disadvantages of OAW include the following:

- Slower welding process than other methods
- Uses volatile and potentially dangerous gases
- Fuel-gas and oxygen cylinders require special handling to avoid damage
- Damaged cylinders can result in fire or explosions

EQUIPMENT

A typical oxyacetylene welding station includes the following components:

- Welding torch and nozzle
- Oxygen cylinder, oxygen regulator, and oxygen hose
- Acetylene cylinder, acetylene supply, acetylene regulator, and acetylene hose
- Flashback arrestors and check valves and torch lighter/spark lighter

Welding Torch

The welding torch is designed to mix oxygen and acetylene in nearly equal amounts and then to ignite and burn the gas mixture at the torch tip. The welding torch has two tubes, one for oxygen and the other for acetylene; a mixing chamber; and oxygen and acetylene valves to control and adjust the flame.

Welding-Torch Tips

Welding tips may be purchased in a wide variety of sizes and shapes. The suitability of a particular welding-tip design depends on a number of factors including the accessibility of the area being welded, the rate of welding speed desired, and the size of the welding flame required for the job.

Manufacturers have their own numbering systems to indicate different welding-tip sizes. There is no industry standard, although comparison charts are available. Keep in mind that tip-size identifications have no bearing on minimum or maximum gas consumption or on flame characteristics. Drill size alone also fails to give an adequate comparison between the various makes of welding tips with identical tip-drill size. The internal torch and tip construction may vary. The gas exit velocities as well as gas-pressure adjustments may also vary.

The welding-torch nozzle is replaceable and is available in a wide variety of sizes. The size selected will depend on the thickness of the metal being welded. Table 14.1 contains data for the selection of welding tips.

Hoses and Hose Connections

Hoses used in oxyacetylene and other gas-welding processes should be strong, nonporous, flexible, and not subject to kinking. The best hoses are constructed from nonblooming neoprene tubing reinforced with braided rayon. The outer coating should be resistant to oil and grease and tough enough to survive most shop

TABLE 14.1   Oxyacetylene welding tips data.

Tip Size	Drill Size	Oxygen Pressure psi		Acetylene Pressure psi		Acetylene		Metal Thickness
		Min.	Max.	Min.	Max.	Min.	Max.	
000	75	¼	2	½	2	½	3	up to 1/32"
00	70	1	2	1	2	1	4	1/64"–3/64"
0	65	1	3	1	3	2	6	1/32"–5/64"
1	60	1	4	1	4	4	8	3/64"–3/32"
2	56	2	5	2	5	7	13	1/16"–1/8"
3	53	3	7	3	7	8	36	⅛"–3/16"
4	49	4	10	4	10	10	41	3/16"–¼"
5	43	5	12	5	15	15	59	¼"–½"
6	36	6	14	6	15	55	127	½"–¾"
7	30	7	16	7	15	78	152	¾"–1¼"
8	29	9	19	8	15	81	160	1¼"–2"
9	28	10	20	9	15	90	166	2"–2½"
10	27	11	22	10	15	100	169	2½"–3"
11	26	13	24	11	15	106	175	3"–3½"
12	25	14	28	12	15	111	211	3½"–4"

conditions. These hoses may be purchased as single hoses (with 1/8- to ½-inch inside diameters) or as twin double-barreled hoses (one line for the oxygen, the other for the fuel gas) with metal binders at the base. Gas hoses can be purchased in continuous lengths up to 300 feet. Standard reel lots are approximately 100 feet.

In addition, cut lengths (with connections) are available in boxed lengths of 12½, 25, and 100 feet. Figure 14.2 shows a picture of an oxyacetylene welding torch kit that would be required to be installed with the oxygen and acetylene cylinders that are purchased or leased.

Hose Inspection and Maintenance

- Regularly inspect the hoses and connections for damage or wear.

- Replace leaky, worn, or damaged hoses. Do not attempt to repair them. A tape-repaired hose is not a safe seal.

- Replace damaged hose connections.

- Always blow out the hoses before welding or cutting.

- Keep as much hose off the floor as possible to protect it from being run over by equipment or

- stepped on.

- Keep hose runs as short as possible to avoid damage.

- Coil excess hose to avoid kinking and tangles in the line.

An oxygen hose is usually black or green (blue in the United Kingdom). The fuel-gas hose is generally red. The color distinctions are standard for safety reasons. For example, using an oxygen hose to carry acetylene could cause a serious accident. No hose should be used for any purpose other than what it has been designed for. A further safety precaution is found in the design of the threading on the hose connections. The fuel-gas hoses have left-handed threaded connections (and a groove around the outside). The oxygen hoses are fitted with right-handed thread connections, with swivel nuts at both ends.

FIGURE 14.2   A typical oxyacetylene cutting/welding torch kit with regulators.

Note: Use the shortest possible hose length between the cylinders and the torch. You will use less oxygen and acetylene with shorter hoses than with longer ones. You will also experience less pressure drop at the torch when using shorter hoses. A metal clamp is used to attach the welding hose to a nipple. A nut on the other end of the nipple is connected to the regulator or torch. Sometimes the shape of the nipple will indicate its use. A bullet-shaped nipple is generally used for oxygen hoses and a nipple with a straight taper for gas-fuel hoses. Another identification method is a groove running around the center of the acetylene nut. This is an indication of a left-hand thread, which will only screw into the acetylene outlet.

Note: Multiple sets of hoses may be connected to a single regulator on a single set of oxyacetylene cylinders only by installing an approved commercially available fitting listed by a nationally recognized testing laboratory. This fitting must be installed on the output side of the regulator and must have an integral shut-off and reverse-flow check valve on each branch.

Pressure Regulators

Oxygen and acetylene pressure regulators are used to control gas pressure. They do this by reducing the high pressure of the gases stored in the cylinders to a working pressure delivered to the torch and by maintaining a constant gas-working pressure during the welding process. There is a regulator for the oxygen and another for the fuel gas. These gas-pressure regulators are connected between the gas cylinder and the hose leading to the torch.

Regulators may differ according to their capacity (ranging from light to high) and according to the type of gas for which they are designed. For example, an oxygen regulator cannot be used for acetylene gas and vice versa. This prevents them from being attached to the wrong cylinder. Another safety feature is found in the hose outlet from the regulator. The outlet connections for oxygen hoses have right-hand threads. For acetylene and other fuel gases the outlet connections have left-hand threads. Both single- and two-stage regulators are available for use in welding systems. Many regulators are constructed with two gauges. One gauge indicates the pressure of the gas in the cylinder (the high-pressure gauge) and the other indicates the working pressure of the gas being delivered to the torch (the low-pressure gauge).

A single-stage regulator requires torch adjustments to maintain a constant working pressure. As the cylinder pressure decreases, the regulator pressure likewise drops, necessitating torch adjustment. In the two-stage regulator, there is automatic compensation for any drop in cylinder pressure. The two-stage regulator is virtually two regulators in one, which operate together to reduce the pressure progressively in two stages instead of one. The first stage in the two-stage regulator serves as a high-pressure reduction chamber. A predetermined pressure is set and maintained by the spring and diaphragm. The gas (at a reduced pressure) then flows into the second of the two stages, which serves as a low-pressure reduction chamber. In this second chamber, pressure control is controlled by an adjustment screw.

Proper regulator maintenance is important for safe and efficient operation. The following points are particularly important to remember:

- The adjusting screw on the regulator must always be released before opening the cylinder valve. Failure to do this results in extreme pressure against the gauge that measures the line pressure and may cause damage to the regulator.

- Gauge-equipped regulators should never be dropped, improperly stored, or in any way subjected to careless handling. The gauges are extremely sensitive instruments and can be easily rendered inoperable.

- Never oil a regulator. Most regulators will have the instruction "use no oil" printed on the face of both gauges.

- All regulator connections should be tight and free from leaks.

Caution: Never adjust an acetylene regulator to allow a discharge greater than 15-psi (103.4-kPa) gauge.

Flashback Arrestors and Check Valves

Flashback arrestors, also called flame traps, and non-return spring-loaded check valves should be installed between the acetylene and oxygen openings in the torch and the matching hoses.

Note: Some torches are designed with integral flashback arrestors and check valves.

Flashback is a potentially dangerous condition caused by the burning of an oxygen-and-fuel-gas mixture in the mixing chamber of the torch handle instead of at the torch tip. If the burning fuel-gas-oxygen mixture passes through the hoses and regulators to the cylinders, it can result in a fire or an explosion leading to serious injury or even death. Among the causes of flashback are:

- Opening the fuel-gas and oxygen cylinder valves and then attempting to light a torch with a blocked tip

- Loose hose connections and/or hose leaks

- Low gas velocity produced by incorrect gas pressure

- Lighting the torch with a failed oxygen or acetylene regulator

Hose check valves prevent the oxygen and fuel gas from crossing over and mixing together in a volatile mixture at the back of the torch's mixing chamber. Flame arrestors, commonly narrow stainless-steel tubes, stop the flame by absorbing its heat and constricting its passage.

Lighters

Oxygen-and-fuel mixtures should not be ignited with a match. A sudden flare-up on ignition could cause the hands or other parts of the body to be burned. To ensure against such danger, the gas mixture should be ignited with a device that provides the required degree of safety for the welder. One such device is a spark lighter, a simple, inexpensive device made of flint and steel. Some spark lighters are equipped with pistol grips and shoot a shower of sparks at the gas flowing from the torch. Others have a rotating flint holder that permits longer use before having to insert a new flint.

Welding Rods

Gas welding rods, or filler rods, are thin strips of metal used to add metal to the weld during the welding process. During welding, the filler rod melts and deposits its metal into the puddle, where it joins with the molten base metal to form a strong weld. Because the composition of the filler rod must be matched as closely as possible to that of the base metal, the selection of the appropriate rod for the job is extremely important. Choosing the wrong filler rod will result in a weak and ineffective weld. Welding rods are available in a variety of sizes and compositions. The sizes range in diameter from 1/16 to 3/8 inch. The cast iron rods are sold in 24-inch lengths. All others are available in 36-inch lengths.

Gas Welding Filler Rods

RG45

- Copper-coated low carbon steel gas welding rod.

- AWS A5.2 Class RG45.

- Use a neutral flame to avoid excess oxidation or carbon pick-up.

- Commonly used to weld ordinary low carbon steel up to ¼-inch thick.

- Recommended where ductility and machinability are most important.

- Produces high quality welds which are ductile and free of porosity.

- Excellent for steel sheet, plates, pipes, castings, and structural shapes.

- No flux required.

RG60

- Low alloy steel gas welding rod.

- AWS A5.2 Class RG60.

- Use a neutral flame to avoid excess oxidation or carbon pick-up.

- Used to produce high tensile strength quality welds on low carbon and low alloy steels such as sheets, plates, pipes of grade A and B analysis and structural shapes.

- Recommended for critical welds that must respond to the same annealing and heat treatment as regular grades of cast steel.

- The high silicon and manganese composition removes impurities from the molten metal thereby eliminating the need for flux.

- RG60 rod is also used as a filler metal in gas tungsten arc welding (GTAW/TIG).

Bare Brass Rod

- Low-fuming bronze gas welding rods, made of copper tin alloy to produce a rod that flows easily, join a variety of metals including cast and malleable iron, galvanized steel, brass copper, and steel.

- Produce strong joints up to 63,000-psi tensile strength.

- Flux required for bare brass rod.

Flux-Coated Brass or Low-Fuming Bronze Rod

- Same characteristics as bare brass rod.

- Flux contained in rod coating.

Flux

Flux is a material used to prevent, dissolve, or facilitate removal of oxides and other undesirable substances that can contaminate the weld. The flux material is fusible and nonmetallic. As a result of the chemical reaction between the flux and the oxide, a slag is formed, which floats to the top of the molten puddle of metal during the welding process. The slag can then be removed from the surface after the weld has cooled.

The following points should be kept in mind when dealing with fluxes:

- The chemical composition of a flux depends on the metal or metals on which it is to be applied.

- Fluxes may be divided into three main categories:
    1. Welding fluxes
    2. Brazing fluxes
    3. Soldering fluxes

- Welding fluxes are divided into gas-welding fluxes and braze-welding fluxes.

- Fluxes are sold as powders, pastes, or liquids (frequently in plastic squeeze bottles). Quantities are available from as small as ½- to 5-pound cans, to jars to 25 pounds, to larger sizes sold in drums.

- Some powdered fluxes may be applied by dipping the heated welding rod into the can. The flux will stick to the rod. The flux powder may also be applied directly to the surface of the base metal. Some powdered fluxes are mixed with alcohol or water and applied to the surface as a paste.

- Paste or liquid fluxes are applied in the form in which they are purchased.

Welding-Torch Tip Cleaners

Welding-torch tips must be cleaned regularly in order to prolong the life of the tip and to provide consistently high performance. Stainless-steel tip cleaners (wires) are available in various diameters to fit different tip orifices. All deposits on the inside of the tip must be removed without enlarging the size of the opening (orifice).

EQUIPMENT SETUP

Basic Assembly

Follow these directions to set up your oxyacetylene-welding equipment:

1. Fasten the oxygen and acetylene cylinders in an upright position to a welding cart, wall, or fixed vertical surface to keep them from falling over. Use a chain or a nonflammable material to fasten them in place. Locate them as close as possible to the welding job but away from open flames.

2. Remove the caps from both cylinders. Examine the cylinder outlet nozzles for stripped threads or a damaged connection seat.

3. Open and close the oxygen cylinder valve very quickly to blow out any loose dust or dirt that may have accumulated in the outlet nozzle. Wipe the connection seat with a dry, clean cloth.

If the dust or dirt is not removed, it could damage the regulator. These unremoved contaminants can also cause incorrect gauge readings.

Caution: Turn your head away when opening the oxygen cylinder valve and make sure that the oxygen stream is not directed toward another worker, a spark, or an open flame. The high pressure of the oxygen stream can cause serious injury to the eyes. Oxygen will also cause a serious fire if ignited by a spark or flame.

Note: The valve opening and the inlet nipple should be shiny and clean inside and outside. This is particularly important for the oxygen cylinder. Oil or grease in the presence of oxygen is flammable or even explosive. Never allow oxygen to contact oil, grease, or other flammable substances.

4. Repeat step 3 for the acetylene cylinder.

5. Connect the oxygen regulator to the oxygen cylinder.

Note: Regulators must be used only with the gas and pressure range for which they are intended and marked. Cylinder-valve outlets and the matching inlet connections on regulators have been designed to minimize, as far as possible, the chances of making incorrect connections.

6. Connect the acetylene regulator to the acetylene cylinder.

7. Connect the green or black oxygen hose to the oxygen regulator. The oxygen hose has a right-hand thread and must be turned clockwise to tighten. Make the connection tight, but avoid overtightening it.

8. Connect the red acetylene hose to the acetylene regulator. The acetylene hose has a left-hand thread and must be turned counterclockwise to tighten. Again, do not overtighten the connection.

9. Charge the oxygen regulator by slowly opening the oxygen cylinder valve. Opening the valve slowly prevents damage to the regulator seat.

Caution: Never face the regulator when opening the cylinder valve. A defective regulator may allow the gas to blow through with enough force to break the gauge glass, resulting in possible injury to anyone standing nearby. Always stand to one side of the regulator when opening the cylinder valve and turn the valve slowly.

10. Open the oxygen-regulator-adjusting screw (the T-handle on the regulator), blow out any dirt or debris in the oxygen hose, and then close it.

11. Repeat steps 9 and 10 for the acetylene regulator and hose.

12. Connect the oxygen hose to the oxygen-needle valve on the torch and the acetylene hose to the acetylene-needle valve. Again, the oxygen hose has a right-hand thread, whereas the acetylene hose has a left-hand one.

13. Close the welding-torch-needle valves and open the oxygen and acetylene cylinder valves.

Adjust the regulators for a normal working pressure and check for leaks at all the connections with soap and water. Soap bubbles will indicate a leak. Tighten the connections with a wrench. If this fails to eliminate the problem, shut off the oxygen and acetylene, check for stripped threads, defective hoses (old hoses become porous), or other damaged parts, and repair or replace as necessary.

Note: Leaks must be repaired before attempting to use the welding equipment. A leak will not only produce wasted gas, but it can also cause a fire or explosion. Always check the equipment for leaks on a periodic basis, not just when the equipment has been initially set up.

Tip Selection

The tip size will depend on the thickness of the metal being welded. Use a tip with a small opening for welding thin sheet metal, and follow these directions:

Note: It is very important to use the correct tip size, with the proper working pressure. If too small a tip is employed, the heat will not be sufficient to fuse the metal to the proper depth. When the tip is too large, the heat is too great, thereby burning holes in the metal.

1. Select the correct tip size for the welding job. The size of the welding tip depends on many factors, including the thickness of the metal, the welding position, and the type of metal being welded. If the tip is too small for the work, too much time is wasted in making the weld and poor fusion is likely to result. If the tip is too large, it is likely to produce poor metal in the weld (due to over oxidation) and a rough-looking job due to lack of control of the flowing metal.

2. Install the tip in the nozzle.

Torch Lighting

Here's how to light the torch:

1. Point the torch tip down and away from your body.

2. Open the oxygen and acetylene cylinder valves and set the working pressure to correspond to the size of tip being used.

Caution: Always face away from the regulator when opening a cylinder valve. A defect in the regulator may cause the gas to blow through, shattering the glass and blowing it into your face. Remember, oxygen and acetylene are charged in the tanks under a high pressure, and if the gas is permitted to come against the regulator suddenly, it may cause some damage to the equipment.

3. Open the acetylene cylinder valve approximately one complete turn and the oxygen all the way. Next, turn the oxygen and acetylene regulator-adjusting valves to the required working pressures.

4. Open the acetylene-needle valve on the torch about one-quarter turn and spark the lighter at the torch tip.

Caution: If you take too long to spark the lighter, acetylene will build up around the torch tip. When the excess acetylene is finally lit, it may cause an explosion, resulting in burns to the hand.

Note: Never use a match to light a torch. This procedure brings your fingers too close to the tip, and the sudden ignition of the acetylene can cause serious burns.

Caution: Make no attempt to relight a torch from the hot metal when welding in an enclosed box, tank, drum, or other small cavity. There may be just enough unburned gas in this confined space to cause an explosion as the acetylene from the tip comes in contact with the hot metal. Instead, move the torch to the open, relight it in the usual manner, and make the necessary adjustments before resuming the weld.

5.   After igniting the acetylene gas, make adjustments at the acetylene valve to permit enough gas to burn to give the proper intensity of flame. If the acetylene flame is accompanied by a lot of smoke, increase the amount of acetylene until the smoke disappears and the flame seems to "jump" off the torch tip.

6.   When the acetylene flame is properly adjusted, slowly open the oxygen valve to allow the air in the line to escape gradually so that it will not blow out the flame.

7.   The torch oxygen valve is then gradually opened until the flame changes from a ragged yellow flame to a perfectly formed bluish cone. This flame is known as a neutral flame and is the torch flame commonly used for most welding.

Adjusting the Flame

The proportions of oxygen and acetylene can be adjusted to produce a neutral, oxidizing, or carburizing flame. Oxyacetylene welding is normally performed using a neutral flame, produced by mixing roughly equal amounts of oxygen and acetylene. (See Figure 14.3.)

Turning Off the Torch

Here's how to turn the torch off:

1.   Close the acetylene-needle valve on the torch first. The acetylene-needle valve is closed first, since shutting off the flow of this gas will immediately extinguish the flame, whereas if the oxygen is shut off first, the acetylene will continue to burn, throwing off smoke and soot.

2.   Close the oxygen-needle valve on the torch.

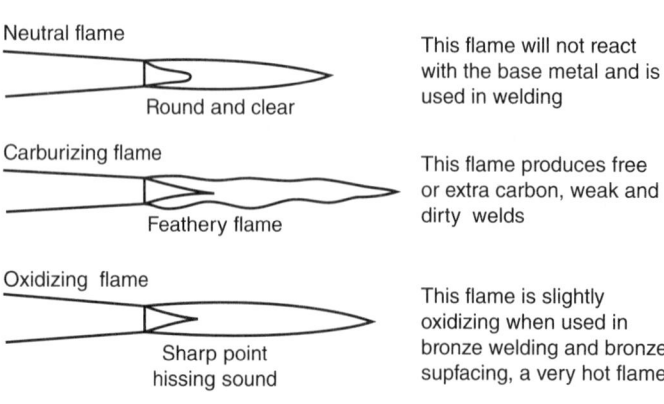

FIGURE 14.3   Oxyacetylene flame tips.

Flame Components

Inner tip is also called the *cone*. That portion of the flame located at the bore of the torch nozzle or the innermost portion of the flame. **Caution:** Never allow the inner tip of the flame to touch the work because it will burn through the metal.

Beard or *brush* is located between the outer envelope and the inner tip or cone. Not always present. **Caution:** Never allow the beard or brush to touch the work.

Outer envelope extends around the beard and inner tip. Much larger in volume than the beard and inner tip because it is fed oxygen from the surrounding atmosphere.

Oxyacetylene Flame Types

Acetylene flame is very white, large, smoky flame produced when torch is first lit.

Carbonizing flame is also called a *reducing flame* and is produced by burning an excess of acetylene. The flame very often has no beard or brush on its inner tip. When adjusted with a small beard, it may be used on most nonferrous metals (those not containing iron elements). The outer envelope is usually the portion of the flame used on these metals.

Neutral flame is produced by burning one part acetylene gas and slightly more than one part oxygen.

Oxidizing flame is produced by burning an excess of oxygen. The flame has no beard and both the inner tip and envelope are shorter. The oxidizing flame is of limited use because it is harmful to many metals.

Equipment Closure

Follow these important steps to shut down your equipment:

1.   If the entire welding unit is to be shut down, shut off both the acetylene and the oxygen cylinder valves.

2.   Remove the pressure on the working gauges by opening the needle valves until the lines are drained. Then promptly close the needle valves.

3.   Release the adjusting screws on the pressure regulators by turning them to the left.

4.   Disconnect the hoses from the torch.

5.   Disconnect the hoses from the regulators.

6.   Remove the regulators from the cylinders and replace the protective caps on the cylinders.

WELDING METHODS

The two basic methods of running a weld bead are the forehand welding method and the backhand welding method. They differ according to whether the torch tip is pointed in the direction of the weld bead or pointed back toward the welded seam. In addition to the forehand and backhand welding methods, the welder is also confronted with the problem of the welding position.

Forehand Method

In the forehand welding method, the tip of the welding torch follows the welding rod in the direction in which the weld is being made (Figure 14.4). This method is characterized by wide semicircular movements

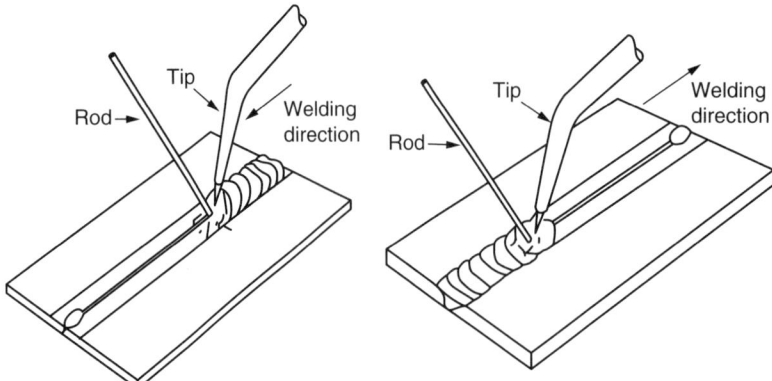

FIGURE 14.4   Oxyacetylene forehand and backhand methods of welding, respectively.

of both the welding tip and the welding rod, which are manipulated so as to produce opposite oscillating movements. The flame is pointed in the direction of the weld but slightly downward so as to preheat the edges of the joint. The major difficulty with the forehand welding method is encountered when welding thicker metals. In order to obtain adequate penetration and proper fusion of the groove surfaces and to permit the movements of the tip and rod, a wide V-groove (90-degree included angle) must be created at the joint. This results in a large puddle, which can prove difficult to control, particularly in the overhead position.

Note: Forehand welding is also sometimes referred to as ripple welding or puddle welding.

Backhand Method

In the backhand welding method, the tip of the torch precedes the welding rod in the direction in which the weld is being made (see Figure 14.4). In contrast to the forehand welding method, the flame is pointed back at the puddle and the welding rod. In addition, the torch is moved steadily down the groove without any oscillating movements. The welding rod, on the other hand, may be moved in circles (within the puddle) or semicircles (back and forth around the puddle).

Backhand welding results in the formation of smaller puddles. A narrower V-groove (30-degree bevel or 60-degree included angle) is required than is the case with the forehand welding method. As a result, greater control is provided as well as reduced welding costs.

Filler Metal

The end of the welding rod should be melted by keeping it beneath the surface of the molten weld puddle. Never allow it to come into contact with the inner cone of the torch's flame. Do not hold the welding rod above the puddle so that the filler metal drips into the puddle.

Torch Angle and Movement

Figures 14.5 and 14.6 illustrate common torch movements, recommended torch angle and distance of the inner cone tip of the flame from the surface, and the weld bead produced by increasing and decreasing the welding speed.

TROUBLESHOOTING

Some of the most common oxyacetylene-welding problems are summarized in Table 14.2.

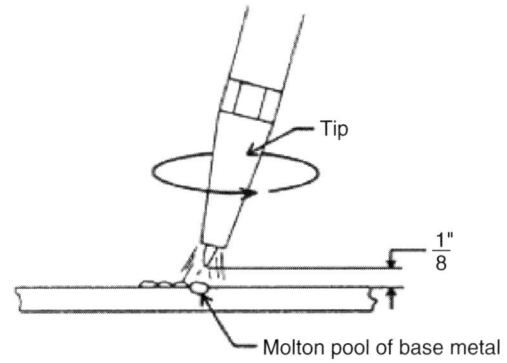

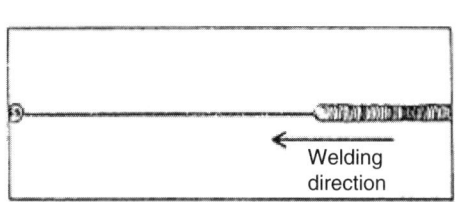

Molton pool of base metal

The tip of the flame is held about 1/8" above the surface of the metal and is moved in a circle.

The tip is moved in a crescent motion, carrying the molten pool of metal across the work to be welded. Filler rod is added in the forward edge of the moltem pool to add reinforcement to the weld.

FIGURE 14.5 Recommended torch angle and inner cone distance from the metal surface.

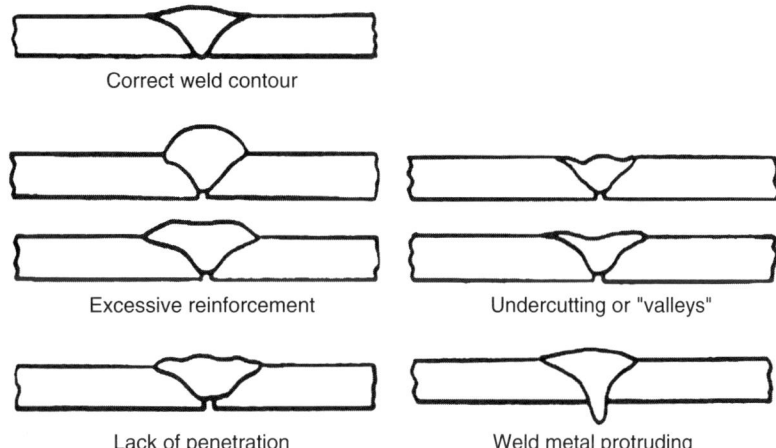

Correct weld contour

Excessive reinforcement

Undercutting or "valleys"

Lack of penetration

Weld metal protruding

FIGURE 14.6 Weld bead produced depending upon increasing and decreasing welding speed.

Note: Do not attempt to re-weld a defective weld. Remove the defective weld metal from the joint and lay down a new weld bead. Re-welding commonly produces weak welds.

TABLE 14.2   Troubleshooting oxyacetylene welding problems.

Problem	Possible Cause	Suggested Remedy
Welding flame splits.	Dirty or clogged torch tip.	Stop welding, allow torch tip time to cool, thoroughly clean the tip, and then resume welding.
Sharp inner cone of flame disappears.	Dirty torch tip.	Stop welding, allow torch tip time to cool, thoroughly clean the tip, and then resume welding.
Premature ignition of gas mixture.	Sparks from weld puddle produce carbon deposits inside nozzle and on torch tip face.	Stop welding, allow torch tip time to cool, thoroughly clean the tip, and then resume welding.
Popping sound, or backfire (A single explosion or pop, or a series of small explosions occurring shortly after the torch is lit. Welding flame may disappear and then reappear, or remain extinguished).	a. Clogged torch tip. b. Pre-ignition of gas mixture in the torch tip, torch mixing chamber, or both. c. Overheated torch tip caused by holding tip to close to the work.	a. Stop welding, allow torch tip time to cool, clean the tip, and then resume welding. b. Increase oxygen and acetylene pressures slowly until popping sound is eliminated. c. Stop welding, allow torch tip time to cool, and then resume welding with tip held farther away from the work.
Squealing sound (Resulting from a very rapid series of small explosions after the torch is lit).	Same as "Popping sound."	Same as "Popping sound."
Flashback (Recession of flame into or in back of the torch mixing chamber).	Equipment problems such as loose or damaged torch tip, clogged torch tip orifices, kinked or damaged hoses; or incorrect welding procedures such as improper gas pressures, over-heated torch tip, or failure to purge hose lines before lighting the torch.	First close the torch oxygen valve and then the acetylene valve. Wait to make sure the fire in the torch or hose has burned out. Check for damage to tip, regulators and hoses, and replace as necessary. Purge the hoses, check oxygen/acetylene pressures, and then relight torch using the standard lighting procedure.
Poor penetration.	Insufficient heat to penetrate to proper depth because torch tip to small.	Replace tip with one of appropriate size.
Holes burned in metal.	Excessive heat caused by tip too large for job.	Replace tip with one of appropriate size.
Uneven weld (Incomplete penetration. Good penetration in spots and partial penetration in between).	Torch is moved along the joint too quickly or slowly, zigzag movements not uniform nor in step with the puddle and zigzag movements overlapping.	Weld with a steady, uniform movement.
Fused portions at one side of the joint and not the other.	Caused by not moving the torch over the joint equally from side to side.	Move the torch so that it touches both sides of the joint uniformly.
Holes in the joint.	Caused by holding the flame too long in one place and overheating the metal.	Move the torch along the joint at a speed guaranteed not to overheat the base metal.

TABLE 14.2   Troubleshooting oxyacetylene welding problems. (*Continued*)

Problem	Possible Cause	Suggested Remedy
Holes in the joint at the end of the weld.	Failure to lift the torch and reduce the heat when reaching the end of the weld.	Use appropriate welding procedure.
Oxide inclusions (Indicated by black specks on the broken surfaces of the weld).	Caused by not adequately cleaning the base metal surface prior to welding.	Make sure the base metal is as clean as possible before welding.
Adhesions	a. Insufficient heat was directed to one side of the joint. b. Welding speed too high causing the weld metal to break cleanly from the metal surface on one side.	a. Apply heat uniformly across joint. b. Decrease welding speed.
Brittle welds	Caused by using a carburizing flame. It is important for the welder to use the correct flame when welding.	Use a neutral flame on ferrous metals.
Overheating	Caused by moving the flame so slowly that too much heat is directed into the weld puddle. This results in the formation of excess metal "icicle" deposits on the bottom of the weld.	Increase torch speed.
Welding (filler) rod sticks in weld puddle.	a. Failure to keep weld puddle molten. b. Failure to dip the welding rod into the weld puddle fast enough.	a. Maintain heat over weld puddle long enough to complete addition of welding rod. b. Dip welding rod into puddle more quickly.

REVIEW QUESTIONS

1. Oxyacetylene welding is performed by the burning of two gases: oxygen and _____.
   a. acetylene
   b. air
   c. hydrogen
   d. nitrogen

2. The pressure of stored acetylene is _____.
   a. 500 psi
   b. 225 psi
   c. 100 psi
   d. 5 psi

3. Acetone is used in a tank of acetylene _____.
   a. to keep the acetylene from exploding
   b. to keep the acetylene gaseous
   c. to keep the acetylene in liquid form
   d. to make the tank of acetylene hard to transport

4. Acetylene tanks are changed in an upright position _____.
   a. for safety purposes
   b. to keep the gas from solidifying
   c. to make it easy to move around
   d. none of these

5. No torch or manifold should have an acetylene pressure above _____ psi for safety.
   a. 5
   b. 10
   c. 15
   d. 25

6. Other gases are supplied in the same type of cylinder as the oxygen cylinder. The gas inside is identified by the _____ of the tank or label.
   a. size
   b. color
   c. weight
   d. height

7. The pressure in the welding torch controlled for use _____.
   a. by a regulator
   b. by a relief plug
   c. by a valve setup
   d. by the atmosphere

8. A _____ flame is used in welding.

   a. neutral

   b. carburizing

   c. oxidizing

   d. none of the above

9. In acetylene welding the tip is moved in a _____ motion, carrying the molten pool of metal across the work to be welded.

   a. circular

   b. straight

   c. crescent

   d. none of the above

10. To shut down after welding, the torch's _____ valve is closed first.

    a. oxygen

    b. acetylene

    c. air

    d. none of the above

11. When the torch is not being used for welding, it should be shut off and placed in its _____.

    a. clamp

    b. holder

    c. rack

    d. none of these

12. If you are unsure of whether or not you have a leak, it is best to:

    a. use the soap test

    b. use a match to see if it lights the gas

    c. use the striker to see if it ignites the gas

    d. go ahead and work as usual

13. If you are setting the pressure in the hoses, with the gas or oxygen flowing into the atmosphere, you must make certain:

    a. that there are no welders in the area

    b. that the tank valve is closed

    c. that there is no open flame or heat source in the area that could ignite the fuel

    d. that there is no leak

14. When the acetylene is first lit, the flame will be:

    a. yellow to orange in color

    b. a black, sooty color

    c. blue with a white tip

    d. white with orange

15. A(n) _____ flame should never be used for welding.

    a. oxidizing

    b. yellow

    c. blue

    d. sooty

16. When finished welding, you should turn off the _____ first.

    a. oxygen

    b. acetylene

    c. air

    d. electricity

17. A pure acetylene flame contains no oxygen and is _____ in color.

    a. white

    b. orange

    c. yellow

    d. blue

18. As oxygen is introduced to an acetylene flame, a cone begins to form, and the flame is called a _____ flame.

    a. carburizing

    b. cutting

    c. welding

    d. neutral

19. When more oxygen is added to the acetylene flame, a _____ flame is the result.

    a. welding

    b. cutting

    c. carburizing

    d. neutral

20. By adding more oxygen to a neutral flame, a _____ flame is produced.

    a. cutting

    b. neutral

    c. carburizing

    d. oxidizing

ANSWERS TO REVIEW QUESTIONS

1. a	2. b	3. a	4. d	5. c
6. b	7. a	8. a	9. c	10. b
11. c	12. a	13. c	14. a	15. a
16. b	17. c	18. a	19. d	20. d

Chapter 15
SHIELDED-METAL-ARC WELDING

Performance Objectives

After studying this chapter you will (be able to):

1. List the advantages and disadvantages of the SMAW process.

2. Describe how the SMAW process works.

3. Explain the equipment that is required for the SMAW process.

4. Differentiate between the various electrodes that are used in the SMAW process.

5. Understand the safety precautions regarding the SMAW process.

Shielded-metal-arc welding (SMAW) is better known as manual-arc welding. It is a portable joining process very adaptable to working on large projects such as buildings, bridges, large tanks, pipelines, and other components or manufactured parts on a site. The work is performed by a welder striking a 14-inch electrode to the work and melting the work pieces together. The melting of metal from the electrode supplies extra metal to produce the bead. With this process, an electric arc is struck between the electrically grounded work and a 9- to 18-inch length of covered metal rod—the electrode. The electrode is clamped in an electrode holder, which is joined by a cable to the power source. The welder grips the insulated handle of the electrode holder and maneuvers the tip of the electrode with respect to the weld joint. When he touches the tip of the electrode against the work and then withdraws it to establish the arc, the welding circuit is completed.

The electrode's coating melts. It vaporizes or breaks down chemically nonmetallic substances incorporated in the covering for the shielding. The mixing of molten base metal and filler metal from the electrode provides the coalescence required to effect joining. As welding progresses, the covered rod becomes shorter and shorter. Finally, the welding must be stopped to remove the stub and replaced with another new electrode.

The periodic changing of electrodes is one of the major disadvantages of the process in production welding. It decreases the operating factor, or the percent of the welder's time spent in the actual operation of laying weld beads. Another disadvantage of SMAW is the limitation placed on the current that can be used. High amperages, such as those used with semiautomatic guns or automatic welding heads, are impractical because of the long (and varying) length of electrode between the arc and the point of electrical contact between the arc and the electrode holder. The welding current is limited by the resistance heating of the electrode. The electrode temperature must not exceed the "break-down" temperature of the covering. If the temperature is too high, the covering chemicals react with each other or with air and therefore do not function properly at the arc. Coverings with organics will break down at lower temperatures than mineral or low hydrogen types of coverings.

The versatility of the process, however—plus the simplicity of the equipment—is viewed by many users whose work would permit some degree of mechanized welding as overriding its inherent disadvantages. This point of view was formerly well taken, but now that semiautomatic self-shielded flux-cored-arc welding has been developed to a simplified (or even superior) degree of versatility and flexibility, there is less justification for adhering to stick-electrode-welding steel fabrication and erection wherever substantial amounts of weld metal must be placed. In fact, the replacement of SMAW with semiautomatic processes has been a primary means by which steel fabricators and erectors have met price squeezes in their welding operations.

Notwithstanding the limitations of SMAW, it is certain to remain a primary welding process. It is one well suited in terms of minimal cost of equipment and broad application possibilities for the home mechanic, the farmer, the repair shop, the garage, the trailer-hitch installer, and many others who are concerned entirely with getting a welding job done.

PRINCIPLES OF OPERATION

Take a look at the basic welding circuit and various others created for the shielded-metal-arc process. Note that welding begins when the arc is struck between the work and the tip of the electrode. The heat of the arc melts the electrode and the surface of the work near the arc. Tiny globules of molten metal form on the tip of the electrode and transfer through the arc into the molten weld "pool" or puddle on the work surface. The transfer through the arc stream is brought about by electrical and magnetic forces. Movement of the arc along the work (or movement of the work under the arc) accomplishes progressive melting and mixing of molten metal, followed by solidification, and thus the unification of parts.

It would be possible to clamp a bare mild-steel electrode into the electrode holder and fuse-join two steel parts. The resulting weld would lack ductility and soundness if judged by present-day standards. The weld metal so deposited would contain oxides and nitrides resulting from reaction of the molten metal with oxygen and nitrogen in the atmosphere. An essential feature of the electrode used in the shielded-metal-arc process is a covering or coating, applied to the core metal by extrusion or dipping, that contains ingredients to shield the arc and protect the hot metal from chemical reaction with constituents of the atmosphere. The shielding ingredients have various functions. One is to shield the arc and provides a dense, impenetrable envelope of vapor or gas around the arc and metal to prevent the pickup of oxygen and nitrogen and the chemical formation of oxides and nitrides in the weld puddle. Another is to provide scavengers and deoxidizers to refine the weld metal. A third is to produce a slag coating over molten globules of metal during their transfer through the arc stream and a slag blanket over the molten puddle and the newly solidified weld. Figure 15.1 shows the decomposition of an electrode covering and the manner in which the arc-stream metal is shielded from the air.

Another function of the shield is to provide the ionization needed for alternating current (AC) welding. With AC the arc goes out 120 times a second. For it to be reignited each time it goes out, an electrically conductive path must be maintained in the arc stream. Potassium compounds in the electrode covering provide ionized gaseous particles that remain ionized during the fraction of a second that the arc is extinguished with AC hertz reversal. An electrical path for re-ignition of the arc is thus maintained. The mechanics of arc shielding vary with the electrode type. Some types of electrodes depend largely on a "disappearing" gaseous shield to protect the arc stream and the weld metal. With these electrodes, only a light covering of slag will be found on the finished weld.

Other electrode types depend largely on slag for shielding. The explanation for the protective action is that the tiny globules of metal being transferred in the arc stream are entirely coated with a thin film of molten slag. Presumably, the globules become coated with slag as vaporized slag condenses on them—so the protective action still arises from gasification. In any event, the slag deposits with these types of electrodes

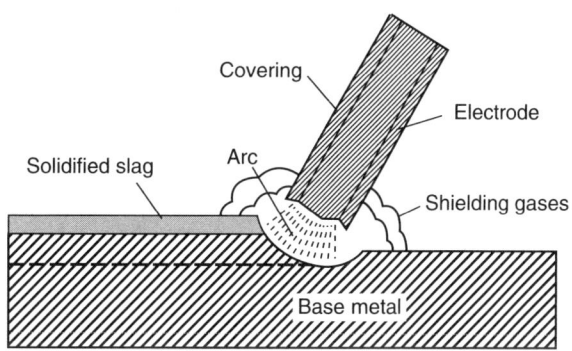

FIGURE 15.1 An SMAW electrode generating a welding bead on a metal plate.

are heavy, completely covering the finished weld. Between these extremes are electrodes that depend on various combinations of gas and slag for shielding.

The performance characteristics of the electrode are related to their slag-forming properties. Electrodes with heavy slag formation have high-deposition rates and are suitable for making large welds downhand. Electrodes that develop a gaseous shield that disappears into the atmosphere and gives a light slag covering are low deposition and best suited for making welds in the vertical or overhead positions.

A solid wire core is the main source of filler metal in electrodes for the shielded-metal-arc process. However, the so-called iron-powder electrodes also supply filler metal from iron powder contained in the electrode covering or within a tubular core wire. Iron powder in the covering increases the efficiency of the arc heat and thus the deposition rate. With thickly covered iron-powder electrodes, it is possible to drag the electrode over the joint without the electrode freezing to the work or shorting out. Even though the heavy covering makes contact with the work, the electrical path through the contained powder particles is not adequate in conductivity to short the arc, and any resistance heating that occurs supplements the heat of the arc in melting the electrode. Because heavily covered iron-powder electrodes can be dragged along the joint, less skill is required in their use.

Some electrodes for the shielded-metal-arc process are fabricated with a tubular wire that contains alloying materials in the core. These are used in producing high-alloy deposits. Just as the conventional electrodes, they have an extruded or dipped covering.

POWER SOURCE

SMAW requires relatively low currents (10–500 amps) and voltages (17–45), depending on the type and size of the electrode used. The current may be either AC or direct current (DC); thus, the power source may be either AC or DC or a combination AC/DC welder. For most work, a variable-voltage power source is preferred, since it is difficult for the welder to hold a constant arc length. With the variable voltage source and the machine set to give a steep volt-ampere curve, the voltage increases or decreases with variations in the arc length to maintain a fairly constant current.

The equipment compensates for the inability of the operator to hold an exact arc length, and he or she is able to obtain a uniform deposition rate. In some welding, however, it may be desirable for the welder to have control over the deposition rate, as when depositing root passes in joints with varying fit-up or in out-of-position work. In these cases variable voltage performance with a flatter voltage-amperage curve is desirable so that the welder can decrease the deposition rate by increasing the arc length or increase it by shortening the arc length. The change from one type of voltage-ampere curve to another is made by changing the open-circuit voltage and current settings of the machine.

The fact that the shielded-metal-arc process can be used with so many electrode types and sizes—in all positions—on a great variety of materials and with flexibility in operator control makes it the most versatile of all welding processes. These advantages are enhanced further by the low cost of equipment. The total advantages of the process, however, must be weighed against the cost per foot of weld when a process is to be selected for a particular job. SMAW is a well-recognized way of getting the job done, but too faithful adherence to it often leads to excessive welding costs. Characteristics of steel electrodes used in SMAW are listed in Table 15.1.

TABLE 15.1 Usability characteristics of mild-steel electrodes.

Type of Electrode	Type of Coating	Welding Position	Type of Current	Penetration	Rate of Deposition	Bead Appearance	Spatter	Slag Removal	Tensile Strength
E6010	High cellulose, sodium	All positions	DC reverse	Deep	Average	Rippled and flat	Moderate	Moderately easy	62,000 psi
E6011	High cellulose, potassium	All positions	AC & DC reverse	Deep	Average	Rippled and flat	Moderate	Moderately easy	62,000 psi
E6012	High titania, sodium	All positions	AC & DC straight	Medium	Good rate	Smooth and convex	Slight	Moderately easy	67,000 psi
E6013	High titania, potassium	All positions	AC & DC reverse	Mild	Good rate	Smooth and flat to convex	Slight	Easy	67,000 psi
E7014	Iron powder, titania	All positions	AC & DC straight	Mild	High rate	Smooth and flat to convex	Very slight	Very easy	72,000 psi
E7016	Low hydrogen, potassium	All positions	AC & DC reverse	Mild to medium	Good rate	Smooth and flat to convex	Slight	Moderately easy	72,000 psi
E6020	High iron oxide	Flat, horz. fillets	AC & DC straight	Deep	High rate	Smooth and flat to convex	Slight	Easy	62,000 psi
E7024	Iron powder, titania	Flat, horz. fillets	AC & DC straight	Deep	Very high rate	Smooth and flat to convex	Very slight	Very easy	72,000 psi
E6027	Iron powder, iron oxide	Flat, horz. fillets	AC & DC straight	Deep	Very high rate	Flat to concave	Slight	Very easy	62,000 psi
E7018	Iron powder, low hydrogen	All positions	AC & DC reverse	High rate	High rate	Smooth and slightly convex	Very slight	Very easy	67,000 psi
E7028	Iron powder, low hydrogen	Flat, horz. fillets	AC & DC reverse	Very high rate	Very high rate	Smooth and slightly convex	Very slight	Very easy	67,000 psi

REVIEW QUESTIONS

1. Shielded-metal-arc welding is better known as _____ arc welding.
    a. automatic
    b. submerged
    c. manual
    d. gas

2. The melting of the electrode supplies the extra metal to produce the _____.
    a. flash
    b. bead
    c. arc
    d. flux

3. The process of welding is dependent primarily on the condition and _____ of the molten pool.
    a. manipulation
    b. temperature
    c. size
    d. color

4. Four factors being controlled by the welder affect the molten pool: length of arc, setting of the power source, angle of the electrode, and _____.
    a. speed of travel
    b. temperature of the pool
    c. size of the pool
    d. angle of the electrode

5. The welding process produces the sound of _____.
    a. frying eggs
    b. birds chirping
    c. clicking
    d. whistling

6. The setting of the power source determines the available _____ for producing the heat.
    a. voltage surge
    b. resistance
    c. wattage
    d. amperage

7. The angle of the electrode is a factor that aids in the formation of the bead and the control of the arc _____.
    a. throw
    b. blow
    c. intensity
    d. brightness

8. The correct angle of the electrode is obtained when it is held vertical to the plate lying on the welding bench and inclined toward the direction of travel from 0 to _____ degrees.
    a. 10
    b. 15
    c. 20
    d. 25

9. The correct speed of travel will produce a bead that is about one-half the rod diameter high and one and one-half times the rod _____ in width.

    a. height
    b. length
    c. diameter
    d. radius

10. The four welding positions are flat, vertical, overhead, and _____.

    a. forward
    b. backward
    c. round
    d. horizontal

11. The two basic types of joints are fillet and _____.

    a. butt
    b. groove
    c. double v
    d. lap

12. In the vertical welding position the welded seam is perpendicular to the earth's _____.

    a. surface
    b. moon
    c. magnetic north pole
    d. magnetic south pole

13. A welder working around a large pipe lying on the ground will have to weld all around the seam, thus all four welding _____ are used.

    a. electrodes
    b. motions
    c. types
    d. positions

14. Which of the following is NOT a function of electrode flux coatings?

    a. makes the arc easy to start
    b. protects the liquid metal from the oxides and nitrides of the atmosphere
    c. stabilizes the arc for heat control
    d. increases splatter during welding

15. Welding rods with heavy coatings of iron powder supply additional amounts of _____ metal to the weld.

    a. slag
    b. filler
    c. fluxed
    d. oxygen

16. Welding rods or electrodes have numbering systems that have been developed by the American Society for Testing Materials (ASTM) and the _____.

    a. National Association of Manufacturers

    b. American Welding Society

    c. American Society of Welding

    d. Welding Society of America

17. Shielded-metal-arc electrodes start with the letter _____.

    a. A   b. E

    c. C   d. D

18. Welding polarity refers to the flow of _____.

    a. current   b. resistance

    c. gas   d. ions

19. In shielded-arc welding, straight polarity has higher melting and deposition rates than other types of _____.

    a. welding   b. resistance

    c. voltage   d. current

20. Reverse polarity delivers maximum penetration under standard welding conditions and gives an advantage for root passes in grooved welding or where fit-up of the parts demands high heat for _____.

    a. fluxing   b. melting

    c. fusion   d. pooling

21. In alternating current the polarity is constantly _____.

    a. reversing   b. failing

    c. breaking down   d. none of the above

22. The size of the electrode and the position of the _____ determine the setting of the welding machine.

    a. welder   b. rod

    c. metal   d. weld

23. Thick metals may require a _____ weld and thus are joined with a series of welds.

    a. special   b. short

    c. long   d. multipass

24. To get maximum penetration with a high _____ rate, two different electrodes may be used on heavy metals.

    a. deposition   b. speed

    c. electrode   d. none of the above

25. Distortion is caused by _____ heat in the work, which in turn produces stress in the weld area.

    a. equal   b. unequal

    c. excess   d. less

ANSWERS TO REVIEW QUESTIONS

1. c	2. b	3. a	4. a	5. a
6. d	7. b	8. d	9. c	10. d
11. b	12. a	13. d	14. d	15. b
16. b	17. b	18. a	19. d	20. c
21. a	22. d	23. d	24. a	25. b

—NOTES—

Chapter 16

GAS METAL ARC WELDING AND FLUX-CORED ARC WELDING

Performance Objectives

After studying this chapter you will (be able to):

1. Explain the advantages and disadvantages of the GMAW process.

2. List the shielding gases used with the GMAW process.

3. Know the variations of the GMAW process.

4. Describe the FCAW process and its similarities with the GMAW process.

5. List the advantages of the FCAW process versus the SMAW process.

Gas metal arc welding (GMAW) is fast and economical because there is no frequent changing of electrodes as with stick-type electrode welding, and there is no slag formed over the weld. The process often can be automated and, if done manually, the welding head is light and compact. GMAW is a logical outgrowth of gas tungsten arc welding. It differs in that the arc is maintained between an automatically fed, consumable wire electrode and the workpiece. It automatically provides the additional filler. It was formerly referred to as metal-inert-gas (MIG) welding.

Argon, helium, or mixtures of the two can be used for welding virtually any metal. These gases are used primarily for welding nonferrous metals. In welding steel, some oxygen (O_2) or carbon dioxide (CO_2) is added to improve the arc stability. It also reduces weld spatter. The cheaper CO_2 alone can be used for welding steel, provided that a deoxidizing electrode wire is used. Shielding gases have a considerable effect on the nature of the metal (drop) transfer from the electrode to the work. They also affect the tendency for undercutting. Electronic controls alter the waveform of the current to make it possible to vary the mechanism of metal transfer by drop spray or shot-circuiting drops.

Some of these variations in the basic process are:

- Pulsed arc welding

- Short-circuiting arc welding

- Spray-transfer welding

- Buried arc welding (CO_2 is used and the arc is buried in its own crater)

GMAW was originally developed to solve the problem with tungsten inert gas (TIG) welding, which could only weld metals up to ¼ inch thick. GMAW or MIG welding can weld metal more than ¼ inch thick. One way in which MIG varies from TIG is that the MIG has an electrode that disappears into the weld puddle, thus producing a strong, clean weld.

Shielding gas for the MIG process is typically:

- CO_2

- Argon/oxygen

- Argon/CO_2

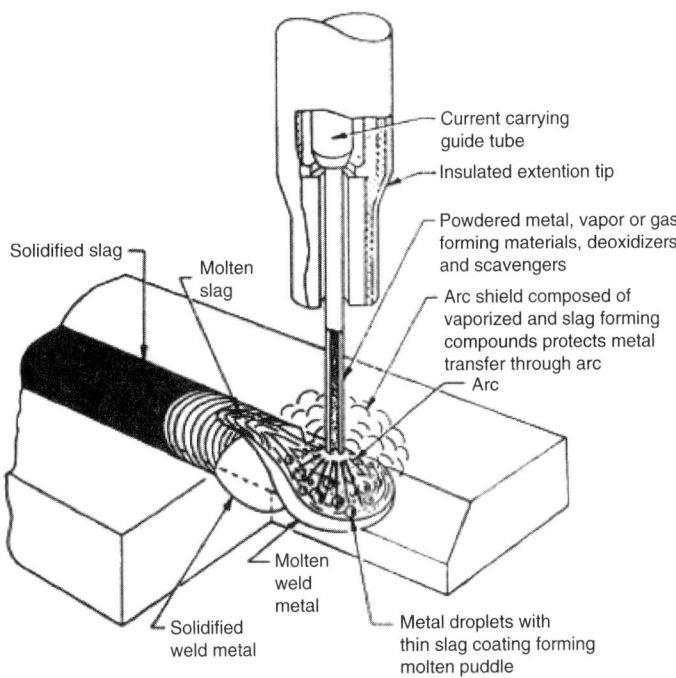

FIGURE 16.1　Flux-cored arc welding (*Lincoln Electric Company*).

The finished weld has no slag and virtually no spatter. This process can be used on a wide variety of material types including:

- Mild steel

- Low-alloy steel

- Aluminum

- Stainless steel

This type of welding is semiautomatic. It uses a handheld gun to which the electrode is fed automatically, or it is used as a full automatic process. The welding guns or heads are similar to those used with gas-shielded flux-cored welding. When the term "manual" is used, it usually refers to the semiautomatic process that utilizes the handheld gun. The self-shielded flux-cored arc welding (FCAW) process came about in the quest for automated welding. The process of putting a 9- or 18-inch electrode into a holder every few minutes takes time and effort on the part of the welder. If the electrode-replacement process could be eliminated, the time lost in changing electrodes would be better utilized in improving productivity. One answer to the problem was to make the electrode of wire so it could be coiled or rolled up and fed out automatically or semiautomatically as needed. It would also be desirable inasmuch as it would allow more metal to be placed in the weld area and thus improve the time used to make a number of passes to accomplish the same result. By putting the flux inside the wire, it was available when needed and did not flake off or crack before it could be used. The outside cover of flux made it easy to handle, and the metal used for the electrode would be smooth on the outside and allow better guidance via a current-carrying guide tube. (See Figure 16.1.)

The welder can activate the system by pressing a trigger to complete the welding circuit. With the semiautomatic gun he or she can reach into areas that are inaccessible to the semiautomatic equipment of other processes.

ADVANTAGES OF THE FLUX-CORED ARC WELDING PROCESS

- When compared with stick-electrode welding, it gives deposition rates up to four times as great, often decreasing welding costs by as much as 50% to 75%.

- It eliminates the need for flux-handling and recovery equipment, as in submerged-arc welding, or for gas, gas-storage, piping, and metering equipment, as in gas-shielded mechanized welding. The semiautomatic process is applicable where other mechanized processes would be too unwieldy.

- It tolerates elements in steel that normally cause weld cracking when stick-electrode or one of the other mechanized welding processes are used.

- It produces crack-free welds in medium-carbon steel using normal welding procedures.

- Under normal conditions, it eliminates the problems of moisture pickup and storage that occur with low-hydrogen electrodes.

- It eliminates stub losses and the time that would be required to change electrodes with the stick-electrode process.

- It eliminates the need for wind shelters, required with gas-shielded welding in field erection, and permits fans and fast air-flow ventilation systems to be used for worker comfort in the shop.

- It permits more seams to be welded in one pass, saving welding time and the time that otherwise would be consumed in between-pass cleaning.

- It is adaptable to a variety of products and permits continuous operation at one welding station, even though a variety of assemblies with widely different joint requirements are run through it.

- It provides for fast filling of gouged-out voids, often required when making repairs to welds or steel castings.

- It gives the speed of mechanized welding to close quarters and reaches into spots inaccessible by other semiautomatic processes.

- It enables the bridging of gaps in fit-up by allowing operator control of the penetration without reducing the quality of the weld, minimizing repair, rework, and rejects.

REVIEW QUESTIONS

1. Another name for gas metal arc welding is:
   - a. tungsten
   - b. submerged
   - c. MIG
   - d. TIG

2. The following gases may be used for proper gas metal arc welding operation:
   - a. helium, carbon dioxide
   - b. helium, argon, carbon dioxide
   - c. air, oxygen, carbon dioxide
   - d. oxygen, argon

3. Although the gas shield is effective in shielding the molten metal from the air, _____ are usually added as alloys in the electrode.
   - a. deoxidizers
   - b. oxidizers
   - c. chemicals
   - d. none of these

4. When the term manual gas metal arc welding is used, the _____ process with its handheld gun is implied.
   - a. automatic
   - b. semiautomatic
   - c. both of these
   - d. neither of these

5. Which technique of welding is meant by the term short welding?
   - a. shielded welding
   - b. gas welding
   - c. spray-arc welding
   - d. short-circuiting welding

6. To use short-arc welding efficiently, special power sources with adjustable slope, voltage, and _____ characteristics are required.
   - a. resistance
   - b. inductance
   - c. capacitance
   - d. current

7. Which type of welding is a logical outgrowth of gas tungsten arc welding?
   - a. pressure-gas welding
   - b. arc welding
   - c. gas tungsten arc welding
   - d. flux-cored arc welding

8. Gas-metal-arc welding was originally developed to solve the problem of welding metals thicker than _____ inch using the TIG method.

   a. ½ inch   b. ¼ inch

   c. 1 inch   d. 1½ inches

9. In the gas-metal-arc-welding process the electrode is consumed; however, in the _____ welding process the electrode is not consumed.

   a. DC arc-   b. tungsten

   c. TIG   d. MIG

10. MIG welding produces a:

    a. glob of weld fill with lots of slag

    b. spotty, slag-covered weld

    c. poor weld but clean

    d. strong, clean weld

11. The self-shielded flux-cored-welding process is an outgrowth of the _____ -metal-arc welding process.

    a. coded   b. shielded

    c. fillet   d. stressed

12. Where is the flux located in the flux-cored-welding process?

    a. on the outside of the electrode

    b. in the middle of the electrode

    c. in the weld puddle

    d. in the arc it produces

13. How long is the electrode in flux-cored welding?

    a. 9 inches   b. 18 inches

    c. an almost unlimited coil   d. 30 inches

14. Why can flux-cored welding be used in industry so economically?

    a. It is easily adapted to automatic operation.

    b. It is semiautomatic in operation.

    c. It is manually operated.

    d. It is electronically controlled.

15. The advantage of semiautomatic operation of the flux-cored process is:

    a. it can reach into areas that are inaccessible to other processes.

    b. it is simple to learn to operate.

    c. it is lightweight.

    d. it is less expensive than other processes.

16. One reason for incorporating the flux inside a tubular wire is to make feasible the coiling of the _____.

    a. flux
    b. electrode holder
    c. electrode
    d. cables

17. What are the advantages of the flux-cored semiautomatic process over the stick-electrode?

    a. It has a higher deposition rate.
    b. It can use higher currents.
    c. It has automatic electrode feed.
    d. All of the above
    e. None of the above

18. Although the American Welding Society calls it self-shielded flux-cored arc welding, it is often referred to as open-arc _____ welding.

    a. vapor
    b. semi
    c. pulse
    d. squirt

19. In many shops the trade name for semiautomatic flux-cored arc welding is _____ welding.

    a. outer-shield
    b. inner-shield
    c. shielded
    d. arcless

20. One of the advantages of flux-cored arc welding is its elimination of the need for handling _____.

    a. electrodes
    b. slag
    c. flux
    d. none of the above

ANSWERS TO REVIEW QUESTIONS

1. c	2. b	3. a	4. b	5. d
6. b	7. c	8. b	9. c	10. d
11. b	12. b	13. c	14. b	15. a
16. c	17. d	18. d	19. b	20. c

—NOTES—

Chapter 17
GAS TUNGSTEN ARC WELDING AND PLASMA ARC CUTTING

Performance Objectives

After studying this chapter you will (be able to):

1. Explain when and why the GTAW process was developed.

2. Describe how the GTAW process functions.

3. Distinguish the differences between the GTAW process and the GMAW process.

4. Know which gases work well with the GTAW process.

5. List the advantages of the plasma arc cutting process over the oxyacetylene cutting process.

TUNGSTEN INERT GAS WELDING

The American Welding Society's definition of gas tungsten arc welding (GTAW) or tungsten inert gas (TIG) welding is "an arc-welding process wherein coalescence is produced by heating with an arc between a tungsten electrode and the work." A filler metal may or may not be used. Shielding is obtained with a gas or a gun mixture. TIG welding was developed in the 1940s for welding magnesium, aluminum, and stainless steel. It is widely used in the aircraft industry. TIG uses a tungsten electrode held in a special holder through which an inert gas is supplied with a sufficient flow to form an inert shield around the arc and the molten pool of metal, thus shielding them from the atmosphere. This process uses argon, helium, or an argon/helium mixture for the shielding. Since the tungsten electrodes often are treated with thorium or zirconium, they provide better current-carrying and electron emission. A high-frequency, high-voltage current usually is superimposed on the regular alternating current (AC) or direct current (DC) welding current to make it easier to start and maintain the arc.

The tungsten electrode is virtually not consumed at arc temperatures in the inert gases, and the arc length remains constant so that the arc is stable and easy to maintain. A clean, strong weld is thus produced without slag. If filler is needed, it must be supplied by a separate wire, as in gas-flame welding. However, in many applications where a close fit exists between the parts being welded, no filler metal may be needed. Continuous fine filler is heated by the passage of an AC current so that it melts as it feeds into the weld puddle just behind the arc. This is as a result of the I2R effect. The deposition rate is several times what can be achieved with a cold wire, and it can be increased further by oscillating the filler wire from side to side when making a wide weld. The hot-wire process cannot be used in the welding of copper or aluminum. This is due to their low resistivities.

Gas tungsten arc welding produces very clean welds, and no special cleaning or slag removal is required because no flux is used. With skillful operators, welds that are scarcely visible often can be made. However, the surfaces to be welded must be clean and free of oil, grease, paint, or rust, because the inert gas does not provide any cleaning or fluxing action.

SPOT WELDING

A variation of the gas-tungsten-arc-welding process is used for making spot welds. The two pieces of metal do not necessarily need to be accessible on both sides of the joint. The basic procedure is shown in Figure 17.1. In this process a modified, vented inert-gas, a tungsten arc gun, and a nozzle are used. The nozzle is pressed

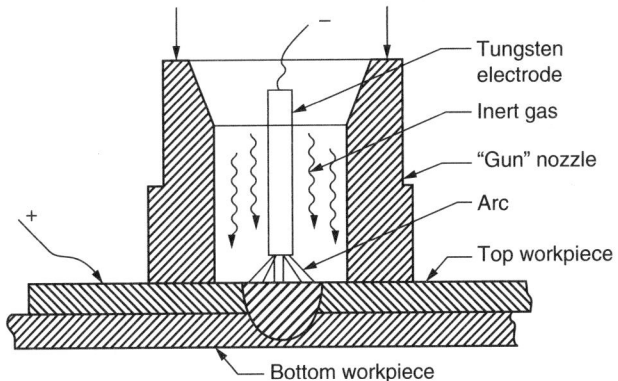

FIGURE 17.1   Making spot welds with the TIG process.

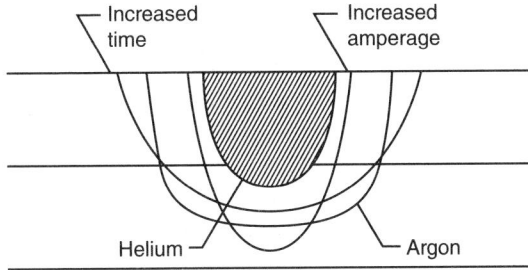

FIGURE 17.2   How changes in time, current, and shielding gas affect the shape of the welding nugget.

against one of the two pieces of the joint. The work pieces must be sufficiently rigid to sustain the pressure that is applied to one side to hold them in reasonably good contact.

The arc between the tungsten electrode and the upper workpiece provides the necessary heat. The inert gas, usually argon or helium, flows through the nozzle and provides a shielding atmosphere. An automatic control moves the electrode to make momentary contact with the workpiece to start the arc. It then withdraws and holds it at a correct distance to maintain the arc. The duration of the arc is timed automatically so that the two workpieces are heated sufficiently to form a spot weld under the pressure of the gun nozzle. The depth and size of the weld nugget are controlled by the amperage, the time, and the type of shielding gas. (See Figure 17.2.) Because access to only one side of the work is required, this type of spot welding has an advantage over resistance spot welding especially in certain applications, as in fastening relatively thin sheet metal to a heavier framework.

PLASMA ARC CUTTING

Plasma arc cutting machines and plasma arc welders are very similar to TIG welders only that they use an extra gas that becomes super-heated into a hot plasma. Torches used in plasma arc cutting produce some of the highest temperatures available from any practicable source. They are very useful for cutting metals, particularly nonferrous and stainless types. Some types of metals cannot be cut by the usual rapid oxidation induced by ordinary flame torches. In addition, plasma arc cutting can cut steel faster and smoother than oxyacetylene cutting. Two types of torches are used in plasma cutting. (See Figure 17.3.) Both of these torches are designed so that the arc column is constricted within a small-diameter nozzle. Through this nozzle passes inert gas. Since the arc fills a substantial part of the nozzle opening, most of the gas flows through the arc and is heated to a very high temperature, thus forming a plasma stream. In the non-transferred type

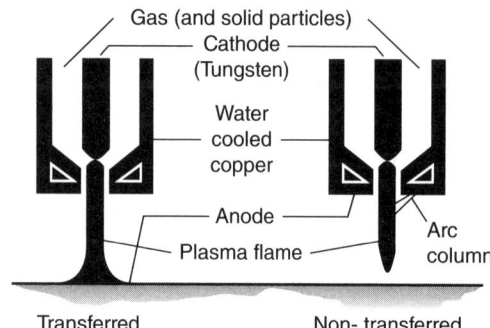

FIGURE 17.3 Principles of plasma arc torches.

of torch, the arc column is completed within the nozzle; it has a temperature range of about 16,640°C or 30,000°F.

With the transferred type of torch the arc column is between the electrode and the work. In this type of torch the temperatures are estimated to be up to 33,316°C or 60,000°F. These high temperatures provide a means for very rapid cutting of any material simply by melting and blowing it from the cut.

Cutting speeds up to 7620 mm or about 300 inches per minute have been obtained in 6.35-mm (¼-inch) aluminum. A combination of extremely high temperature and the jet-like action of the plasma produces narrow kerfs and remarkably smooth surfaces. They are nearly as smooth as you can obtain by sawing. Transferred torches usually are used for cutting metals. The non-transferred type is used for nonmetals.

Gases used in this type of cutting are argon, helium, nitrogen, and mixtures of argon and hydrogen. Mixtures of 65% to 80% argon and 20% to 35% hydrogen are most often used.

A limited amount of welding is done with plasma torches. That is because the temperatures are too high for most work. Good results have been achieved, however, with a torch that integrates a small non-transferred arc within the torch. This heats the orifice gas and ionizes it. The gas forms a conductive path for the main transferred arc. This then permits instant ignition of a low-current arc, which can be lower in magnitude, more stable, and more readily controlled than an ordinary plasma torch. Separate DC power supplies are used for the pilot and main arcs. An inert shielding gas usually is supplied through an outer cup surrounding the torch.

REVIEW QUESTIONS

1. Tungsten inert gas welding is called _____ in the trade.
   a. MIGW
   b. TIGW
   c. MIG
   d. TIG

2. What is the result of the atmospheric contaminants becoming part of the welding process?
   a. clean joints
   b. strong joints
   c. joints with nice lines
   d. weak spots in the joint

3. An inert gas is:
   a. active
   b. inactive
   c. lazy
   d. not reactive with other gases and metals

4. Tungsten inert gas welding was originally developed specifically for welding manganese, aluminum, and _____.
   a. stainless steel
   b. ferrous metals
   c. heavy metals
   d. copper

5. Before TIG welding was introduced, welders had to resort to using _____ to remove contaminants from the weld.
   a. electrodes
   b. heat
   c. skill
   d. flux

6. Tungsten is used for the electrode in TIG welding because:
   a. it is plentiful
   b. it is inexpensive
   c. it does not melt easily
   d. it melts easily

7. The TIG process is particularly suitable for welding _____ materials where the requirements for quality and finish are exacting.
   a. heavy
   b. thick
   c. thin
   d. rough

8. The TIG electrode that does not melt in the arc is made of:
   a. aluminum
   b. copper
   c. tungsten
   d. magnesium

9. The advantage of using a hot wire with oscillation for filler in TIG welding is that:

    a. it has a less rapid deposition rate

    b. it has a more rapid deposition rate

    c. it has a very easy-to-use coating

    d. it can be easily coiled

10. Gas tungsten arc spot welding can be used where only one side of the workpiece is available for applying _____.

    a. pressure   b. heat

    c. cooling water   d. the arc

11. An arc is an electric current flowing between two electrodes through an ionized column of gas called a _____.

    a. flame   b. light source

    c. heat source   d. plasma

12. The arc column is a mixture of neutral and excited gas _____.

    a. current   b. atoms

    c. ions   d. electrons

13. In the central column of the plasma, electrons, atoms, and ions are in accelerated motion and constantly _____.

    a. glowing   b. moving

    c. colliding   d. standing still

14. The outer portion of the arc flame is somewhat _____ and consists of recombining gas molecules that were disassociated in the central column.

    a. warmer   b. hotter

    c. cooler   d. none of the above

15. In welding, the arc not only provides the heat needed to melt the electrode and the base metal but must under certain conditions also supply the means to transport the molten metal from the tip of the electrode to the _____.

    a. work   b. arc stream

    c. heat source   d. none of the above

16. Typical thermal efficiencies for metal-arc welding are in the range of:

    a. 75% to 80%   b. 50% to 60%

    c. 10% to 20%   d. 85% to 95%

17. Typical thermal efficiencies for welding with non-consumable electrodes are in the range of:

    a. 50% to 60%   b. 75% to 80%

    c. 85% to 95%   d. 10% to 20%

18. Arc welding may be done with either AC or DC current and with the _____ either positive or negative.

    a. electrode   b. electrode holder

    c. top piece of work   d. none of the above

19. A limited amount of welding is done with plasma torches, because the temperature is too _____ for most work.

    a. low   b. high

    c. expensive   d. dangerous

20. Good results have been achieved with a plasma torch that integrates a small non-transferred arc within the torch. This heats the orifice gas, ionizing it and thereby forming a conductive path for the main transferred _____.

    a. current   b. electrode

    c. heat source   d. arc

ANSWERS TO REVIEW QUESTIONS

1. d	2. d	3. d	4. a	5. d
6. c	7. c	8. c	9. b	10. a
11. d	12. b	13. c	14. c	15. a
16. a	17. a	18. a	19. b	20. d

—NOTES—

Chapter 18
SUBMERGED-ARC WELDING

Performance Objectives

After studying this chapter you will (be able to):

1. Describe what shields the weld in the submerged-arc-welding process.

2. Explain the importance of the submerged-arc-welding process.

3. Know which metals are suitable for use with the submerged-arc-welding process.

4. List and describe some of the special submerged-arc-welding processes.

Submerged-arc welding is an automatic (or semiautomatic) welding process. It uses a granular flux blanket to completely cover the weld area, while a consumable electrode is continuously and mechanically fed into the arc. (See Figure 18.1.) The name comes from the fact that the arc is hidden or submerged beneath the flux blanket. It is not visible to the welder. This welding process is also referred to as:

- Flux-covered arc welding

- Hidden-arc welding

- Sub-arc welding

- Submerged-melt welding

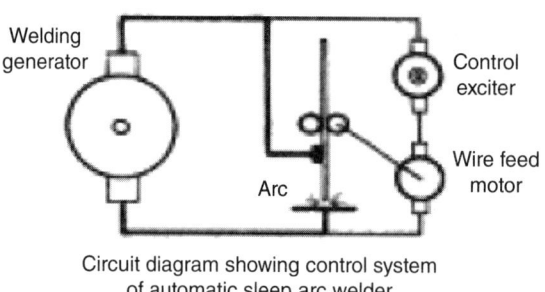

Circuit diagram showing control system
of automatic sleep arc welder

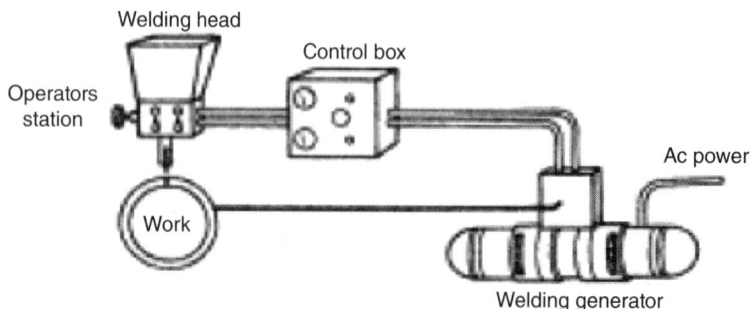

FIGURE 18.1   Submerged-arc-welding process diagram.

Union-melt welding is a submerged-arc-welding process developed by the Linde Air Products Company, Unionmelt® is the trade name for the company's specially developed flux. This is an electric-furnace product carefully controlled as to composition and preparation. Linde is a subsidiary of Union Carbide. Physical characteristics of welds made by the submerged-arc-welding process are such that all types of inspection requirements can be met if the procedures specified are followed. Inspections include:

- Full tensile

- Reduced tensile

- Free bends

- Nick bends

- X-rays

Basic principles of this welding process center on the concept of a gaseous shield around the arc. Slag over the weld protects both the arc and the molten metal from atmospheric contaminants.

Results of this double shielding include deeper penetration, higher welding currents, faster welding, higher weld quality, and lower welding costs. Instead of flux-coated electrodes, granular flux is automatically fed in the latest development; it is in fact an automatic shielded-metal-arc-welding process. Granular flux is deposited on the joint to be welded and fed deep enough to cover the completed weld. A bare metal welding-electrode wire is fed into the blanket of flux. The rate of feed is controlled automatically for proper arc length. Direct current (DC) produces the arc between the electrode wire and the joint. The resultant heat from the arc fuses electrode and base metal, producing the weld.

Flux adjacent to the arc melts, floats on the surface of the molten metal, then solidifies as a slag on top of the weld. Since the arc and molten metal are blanketed by flux at all times, the metal is completely protected from contact with the air, assuming maximum quality of welds and making possible the use of very high amperage for faster welding. (See Figure 18.2.)

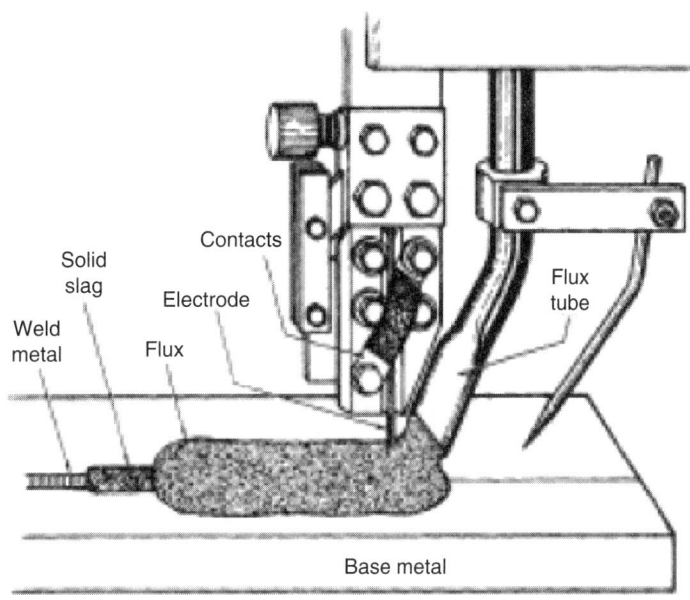

FIGURE 18.2 Distribution of flux from the hopper.

SUBMERGED-ARC-WELDING EQUIPMENT

A submerged-arc-welding system generally consists of the following basic components:

1. An AC or DC power source

2. A welding rod

3. An electrode reel and motor to feed the electrode wire into the arc

4. A flux hopper and a unit to recover unfused flux

5. A control unit

Almost any kind of standard DC generator can be used for submerged-arc welding. The heavy-duty AC transformers used in this welding process usually have a 1000-amp capacity (1000 amps for 1 hour of operation, 750 amps of continuous operation). However, both smaller and larger units are also used. Current higher than that obtainable from one transformer may be had by connecting two or more units in parallel. DC generators may also be connected in parallel to obtain sufficient amperage. (See Figure 18.3.) Both AC and DC power sources must use a means of remote control for the welding current, so that welding can be started and stopped by a switch mounted near the welding head. This is provided by the control unit. If AC is used, a magnetically operated contactor or heavy-duty circuit breaker should be installed in the primary supply of the transformer. The supply leads should have automatic cutout protection of sufficient capacity to protect the primary supply. The control unit also indicates voltage and amperage levels and provides controls for current adjustments, the rate of electrode speed, and travel-speed adjustments.

WELDING CURRENT

Either DC or single-phase AC may be used for submerged-arc welding. AC is the more frequently used. An open-circuit welding of 60 to 100 volts can be used. Table 18.1 gives approximate maximum currents

FIGURE 18.3 Submerged-arc-welding head with wire spool and flux hopper.

TABLE 18.1 Current ranges for submerged-arc-welding electrodes.

Wire Diameter (in)	Current (amp)
3/64	200–600
3/32	230–700
⅛	300–900
3/32	420–1000
3/16	480–1100
7/32	600–1200
¼	700–1600
5/16	1000–2500
⅜	1500–4000

required for various thicknesses of one-pass for Unionmelt® butt-welds and fillet welds. Sufficient adjustable reactance or resistance should be provided, if using DC, to permit continuous step-less adjustment and control of the amperage during the welding operation. The reactance (or resistance in DC circuits) may be built into the power-supply unit or be connected separately. Actual voltage at the welding zone will be adjusted through the special voltage control at approximately 25 to 50 volts, depending on the size and shape of the weld.

Submerged-arc welding uses a higher current than that used in the manual arc-welding processes. This results in such welding characteristics as:

- Reduced weld shrinkage

- Minimum distortion of the welded structure

- Faster filler-metal deposition rate

- Faster welding speed

FLUXES AND ELECTRODES

Fluxes used in submerged-arc welding must not produce large amounts of gas. These fluxes are granular fusible substances, available in a number of different grades. Each grade differs somewhat in chemical composition. The selection of an appropriate flux depends on a number of factors, including:

- Type of metal to be welded

- Welding speed

- Thickness of the metal

Flux is fed around the electrode during the welding process. Use just enough flux to submerge the arc. However, there will be no harmful effect on the weld if the arc occasionally breaks out of the flux. A special aspect of submerged-arc welding is that alloys can be added to the weld through the flux. A mild-steel electrode in conjunction with a special agglomerated-alloy flux is used to do this. Fluxing and alloy materials are ground and mixed together in an agglomeration process so that neither can be separated from the other. The result is an agglomerated-alloy flux that will ensure stable, uniform welds. A wide variety of fluxes are

TABLE 18.2   Typical mechanical electrode strength.

AWS Classification Flus Wire	Tensile Strength (100 psi)	Yield Strength (1000 psi)	Impact Strength, Charpy v-(ft-lb)		Elongation in 2 inch (5)
			at 0°F	at −20°F	
	74	60	19	—	28
	73	60	50	—	26
	77	61	24	—	28
	71	56	—	61	30
	74	58	—	21	28
	77	64	—	42	26
	68	57	—	45	20
	78	66	21	18	25
	79	68	25	22	27
	89	77	50	25	25
	85	70	22	—	25
	83	75	—	47	26
	86	74	—	40	27
	85	72	—	27	24

possible by varying the alloy content. Because of the alloying elements that can be incorporated in the flux, the submerged-arc-welding process can be applied to a wide range of jobs using only a small selection of standard electrodes.

Fluxes used in submerged-arc welding are not classified. Therefore, the welder must follow the manufacturer's recommendations for a particular job. Fluxes used in this process are relatively inexpensive. They should be stored in a dry place. If the flux becomes wet, dry it out by heating it to 500 or 600°F. Fused fluxes may be ground up and reused if they are mixed with at least 75% of fresh flux. Grinding costs generally make this practice uneconomical.

Instead of stick electrodes, the submerged-arc-welding process uses coils of bare metal wire that are continuously and mechanically fed into the arc. In this respect, submerged-arc welding strongly resembles MIG welding, but can deposit a lot more weld in one pass. The electrode wire diameters generally range from 1/5 to ¼ inch and may be purchased in a number of different coil sizes: 1, 5, 10, 25, and 35 pounds and up.

JOINT PREPARATION

Tables 18.1 and 18.2 illustrate equipment settings for submerged-arc welding. On a curved surface, such as small-diameter-girth welds, the speeds and currents are less than the equivalent seam on a flat, horizontal surface. The reason for this is that the molten flux and steel are very fluid and it is necessary to limit the amount of molten material so that it does not run off the curved surface. To control this tendency to spill off, the current and the speed are reduced and the point of welding is ½ to 2½ inches off the vertical centerline in the direction opposite the rotation of the work. (See Figure 18.4.) Where no gap is specified, the seam should be fitted tightly together. In butt seams, if the gap is 1/32 inch or more due to poor fit-up, the seam must be sealed with a sealing bead.

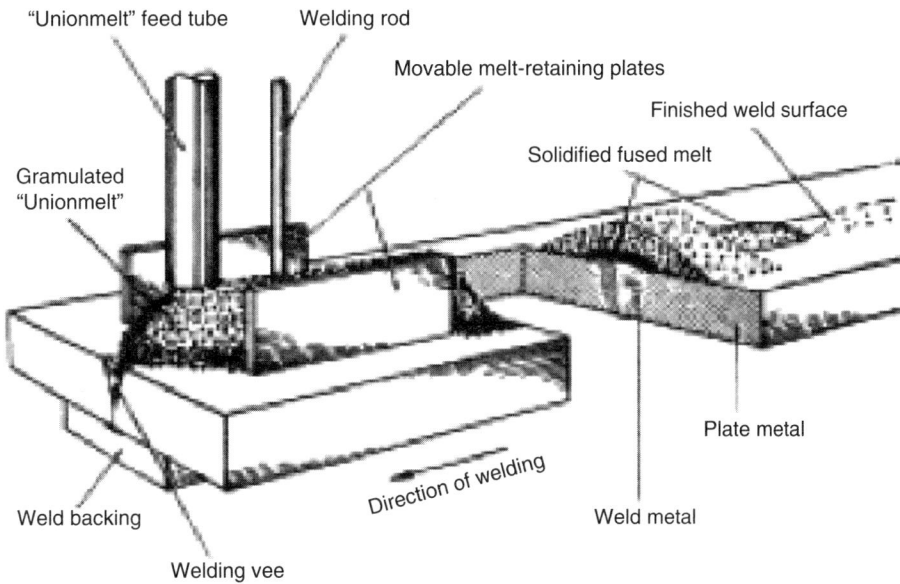

FIGURE 18.4   Cutaway view of the submerged-arc-welding bead.

SPECIAL SUBMERGED-ARC-WELDING PROCESSES

Special jobs or application can call for any of the several variations of the submerged-arc-welding process. They have been developed to answer the needs of these special applications. For the most part, these variations are multiple-arc-welding procedures, including:

- Union-melt arc welding
- Twin-arc welding
- Tandem-arc welding
- Three-o'clock welding
- Series-arc welding
- Arc-strip welding

Union-melt welding has been discussed previously. Twin-arc welding is a submerged-arc-welding process that consists of two electrodes fed through the same welding head for a fast single-pass method of welding large fillet welds in the flat position and in wide V-joints. DC current is recommended. Tandem-arc welding employs two or more electrodes used in tandem (one following the other). This will allow for large deposition rates in one pass. Each electrode will have its own separate power source. Either AC or DC can be used, often in combination.

Three-o'clock welding (developed by the Lincoln Electric Company) is a submerged-arc-welding process in which the joint can be placed in the horizontal position without loss of flux or weld metal. The electrode is directed into the joint. Series-arc welding (developed by Union Carbide and Carbon Research Labs) involves the use of two converging electrodes connected in series with a suitable power source for the purpose of making welds with a very shallow penetration. Arc-strip welding (developed by the Arcos Corporation) is a submerged-welding process in which a 2-inch-wide flat stainless-steel strip is substituted for the bare metal wire used by other processes.

SUBMERGED-ARC-WELDING APPLICATIONS

The submerged-arc-welding process is especially suitable for welding soft metals such as:

- Copper and copper alloys
- Nickel and nickel alloys
- Low- and mild-alloy steels
- Stainless steel

This welding process is done in the flat and horizontal positions, which limits its application. Submerged-arc welding is basically an automatic process in which either the equipment or the work may be moved during the welding. Submerged-arc welding is adaptable to both automatic and semiautomatic welding methods. Both the direction and travel speed of the welding head are controlled by hand. Both procedures can be applied to a vast number of welding jobs ranging from the relatively minor operations of welding axles and spindles to such large-scale ones as welding girders in major construction projects. Enough accessory equipment is available in submerged-arc welding to make it an extremely flexible and adaptable welding method. Submerged-arc welding has become almost synonymous with automatic and semiautomatic welding.

REVIEW QUESTIONS

1. Submerged-arc welding is an automatic welding process that uses a _____-flux blanket to completely cover the weld area while a consumable electrode is continuously and mechanically fed into the arc.

   a. wet
   b. liquid
   c. dry
   d. granular

2. Basic principles of submerged-arc welding center on the concept of a _____ shield around the arc.

   a. wooden
   b. metal
   c. gaseous
   d. none of the above

3. The rate of wire feed is controlled _____ for proper arc length.

   a. automatically
   b. manually
   c. by the arc itself
   d. none of the above

4. Flux adjacent to the arc floats on the surface of the molten metal and solidifies as _____ on top of the weld.

   a. glass
   b. slag
   c. flux
   d. none of the above

5. Submerged-arc welding uses a higher _____ than that used in the manual arc-welding processes.

   a. current
   b. voltage
   c. resistance
   d. reactance

6. Which of the following is not a characteristic of submerged-arc welding?

   a. reduced weld shrinkage
   b. minimum distortion of the welded structure
   c. a faster welding speed
   d. an easily broken weld joint

7. Selection of an appropriate flux depends on which of these factors?

    a. type of metal to be welded

    b. weather to which joint is to be subjected

    c. age of the metal being welded

    d. none of the above

8. Fluxes used in submerged-arc welding are relatively _____.

    a. inexpensive   b. expensive

    c. rare   d. common

9. Instead of stick electrodes, the submerged-arc-welding process uses coils of _____ wire.

    a. coated   b. bare

    c. kinked   d. tungsten

10. Submerged-arc welding is adaptable to both _____ and semiautomatic welding methods.

    a. automatic   b. manual

    c. robotic   d. none of the above

ANSWERS TO REVIEW QUESTIONS

1. d	2. c	3. a	4. b	5. a
6. d	7. a	8. a	9. b	10. a

GLOSSARY

alloy pipe

A steel pipe with one or more elements other than carbon which gives it greater resistance to corrosion and more strength than carbon steel pipe.

angle of bend

In a pipe, the angle at the center of the bend between radial lines from the beginning and end of the bend to the center.

angle valve

A valve, usually of the type in which the inlet and outlet are at right angles.

backing ring

A metal strip used to prevent melted metal from the welding process from entering a pipe when making a butt welded joint.

bell-and-spigot joint

The commonly used joint in cast iron pipe. Each piece is made with an enlarged diameter or bell at one end into which the plain or spigot end of another piece is inserted when laying. The joint is then made tight by cement, oakum, lead, or rubber caulked into the bell around the spigot.

black pipe

Steel pipe that has not been galvanized.

blank flange

A flange in which the bolt holes have not been drilled.

blind flange

A flange used to seal off the end of a pipe.

bonnet

Part of a valve used to guide and support the valve stem.

branch

The outlet or inlet of a fitting not in line with the run and taking off at an angle with the run.

branch tee

A tee having many side branches.

brazed

Joined by hard solder.

bullhead tee

A tee the branch of which is larger than the run.

bushing

A pipe fitting for connecting a pipe with a female fitting of larger size. It is a hollow plug with internal and external threads.

butt-weld joint

A welded pipe joint made with the ends of the two pipes butting each other, the weld being around the periphery.

butt-weld pipe

A pipe welded along a seam butted edge to edge and not scarfed or lapped.

bypass

In a pipeline, a supplementary line leaving the main run and rejoining it at some point beyond a valve or other apparatus, so that service is not interrupted when the valve or apparatus is not usable.

carbon steel pipe

Steel pipe that owes its properties chiefly to the carbon that it contains.

check valve

A valve designed to allow a fluid to pass through in one direction only. A common type has a plate that swings open when the flow of a liquid presses upon it. The reverse flow is aided by gravity in forcing the plate to swing back against a seat, thereby shutting off the reverse flow.

close nipple

A nipple with a length twice that of standard pipe thread.

companion flange

A pipe flange used to connect with another flange or with a flanged valve or fitting. It is attached to the pipe by threads, welding, or other method and differs from a flange that is an integral part of a pipe or fitting.

compression joint

A multipiece joint with cup-shaped threaded nuts that, when tightened, compress tapered sleeves so that they form a tight joint on the periphery of the tubing they connect.

condensate

When steam loses sufficient heat, it returns to water; this liquefied steam is condensate.

coupling

A threaded sleeve used to connect two pipes. They have internal threads at both ends to fit extended threads on pipe.

cross

A pipe fitting with four branches in pairs, each pair on one axis; the axes are at right angles.

cross valve

A valve fitted on a transverse pipe such that open communication is possible between two parallel pipes.

crossover

A small fitting that has a double offset or is shaped like the letter U with the ends unturned out. It is only made in small sizes and is used to pass the flow of one pipe past another when the pipes are in the same plane.

cup weld

A pipe weld where one pipe is expanded on the end to allow the entrance of the end of the other pipe. The weld is then circumferential or cup-shaped at the end of the expanded pipe.

double elbow

A small ell used in a gas fitting. These fittings have wings cast on each side, and the wings have holes so that they may be fastened by wood screws to a ceiling, wall, or framing timbers.

double extra-strong pipe

A schedule of steel or wrought iron pipe that typically has a wall thickness twice that of extra-strong pipe.

double sweep tee

A tee made with a gradual, long radius curves between body and branch.

double tee

A tee having the wings the same as the drop elbow.

elbow

A fitting that makes an angle between adjacent pipes. The angle is 90 degrees, unless another angle is specified.

expansion joint

A joint whose primary purpose is not to join pipe but to absorb that longitudinal expansion in the pipe line due to heat.

expansion loop

A large radius bend in a pipe line to absorb longitudinal expansion in the line due to heat.

flange

In pipe work, a ring-shaped plate on the end of a pipe at right angles to the end of the pipe and provided with holes for bolts to allow fastening the pipe to a similarly equipped adjoining pipe. The resulting joint is a flanged joint.

flange faces

Pipe flanges that have the entire face of the flange faced straight across and use either a frill face or ring gasket. They are commonly employed for pressures less than 125 pounds on steam and water lines.

fusion weld

Joining metals by fusion, using oxyacetylene or electric arc.

galvanized pipe

Steel pipe coated with zinc to resist corrosion.

gate valve

A valve employing a gate, often wedge-shaped, allowing fluid to flow when the gate is lifted from its seat. Such valves have less resistance to flow than globe valves.

globe valve

A valve with a somewhat globe-shaped body with a manually raised or lowered disk that when closed rests on a seat so as to prevent passage of a fluid.

ground joint

A joint where the parts to be joined are precisely finished and then ground in so that the seal is tight.

header

A large pipe or drum into which each of a group of boilers is connected. Also, used for a large pipe from which a number of smaller ones are connected in line and from the side of the large pipe.

lap weld

Made by welding along a "scarfed" longitudinal seam in which one part is overlapped by the other.

lapped joint

A pipe joint made by using loose flanges on lengths of pipe whose ends are turned over or lapped over to produce a bearing surface for a gasket or metal-to-metal joint.

lead joint

A joint made by pouring molten lead into the space between a bell and spigot and making the lead tight by caulking.

lip union

A form of pipe union characterized by the lip that prevents the gasket from being squeezed into the pipe so as to obstruct the flow.

malleable iron

Cast iron that is heat-treated to reduce brittleness. The process enables the material to stretch to some extent and to withstand greater shock.

manifold

A fitting with a number of branches in line connecting to smaller pipes. Used largely as an interchangeable term with *header*.

medium pressure

When applied to valves and fittings, implies they are suitable for a working pressure from 125 to 175 pounds per square inch (psi).

mill length

Also known as random length. Run-of-mill pipe is 16 to 20 feet in length. Some pipe is made in double lengths of 30 to 35 feet.

needle valve

A valve provided with a long tapering point in place of the ordinary valve disk. The tapering point permits fine graduation of the opening.

nipple

A tubular pipe fitting usually threaded on both ends and under 12 inches in length. Pipe over 12 inches long is regarded as cut pipe.

O.D. pipe

Pipe designated by its outside diameter, or O.D. This designation refers to the nominal size of pipe with an outside diameter over 14 inches. Pipe with a smaller diameter is usually designated by its inside diameter (I.D.).

offset

When, to avoid an obstruction or for some other reason, a line of pipe is to be continued as a line running in the same direction, the perpendicular distance between the centerline of the one and the extension of the other is the offset. Travel is the length of the connecting line measured between the length of the centerline intersections, and run is the other leg of the triangle formed by the travel and the offset.

plug valve

A valve with a short section of a cone or tapered plug through which a hole is cut so that fluid can flow through when the hole aligns with the inlet and outlet, but flow is blocked when the plug is rotated 90 degrees.

reducer

A fitting with a larger size at one end than the other; the larger size is designated first. Reducers are threaded inside, unless specified as flanged or welded or for some special joint.

relief valve

A valve designed to open automatically to relieve excess pressure.

resistance weld pipe

Pipe made by bending a plate into circular form and passing electric current through the material to obtain a welding heat.

rolling offset

Same as offset, but used where two lines are not in the same vertical or horizontal plane.

rotary pressure joint

A joint for connecting a pipe under pressure to a rotating machine.

run

A length of pipe made of more than one piece of pipe or a portion of a fitting having its ends in a straight line or nearly so.

saddle flange

A flange curved to fit a boiler or tank and to be attached to a threaded pipe. The flange is riveted or welded to the boiler or tank.

saturated steam

Steam at the same temperature at which water boils and under the same pressure.

screwed flange

A flange screwed on the pipe that it is connecting to an adjoining pipe.

screwed joint

A pipe joint consisting of threaded male and female parts screwed together.

seamless pipe

Pipe or tube formed by piercing a billet of steel and then rolling.

service fitting

A straight ell or straight tee that has external threads at one end and internal threads at the other.

service pipe

A pipe that connects water or gas mains to a building.

set

Same as an offset, but also used in place of an offset where the connected pipes are not in the same vertical or horizontal plane—a rolling offset.

setback

In a pipe bend, the distance measured back from the intersection of the centerlines to the beginning of the bend.

short nipple

A nipple whose length is a little greater than that of two threaded lengths or somewhat longer than a close nipple, so that it has some threaded portion between the two threads.

shoulder nipple

A nipple of any length that has a portion of pipe between two pipe threads. Generally used, however, it is a nipple halfway between the length of a close nipple and a short nipple.

sleeve weld

A joint made by butting two pipes together and welding a sleeve over the outside.

slip-on flange

A flange slipped over the end of the pipe and then welded to the pipe.

socket weld

A joint made by use of a socket weld fitting that has a female end or socket for insertion of the pipe to which it is welded.

solder joint

A method of joining pipe and/or fittings by use of solder applied with heat.

spiral pipe

A pipe made by coiling a plate into a helix and riveting or welding overlapped edges.

stainless-steel pipe

An alloy steel pipe with corrosion-resisting properties, usually imported by nickel and chromium plating services.

standard pressure

Formerly used to designate cast iron flanges, fittings, valves, etc., suitable for a maximum working steam pressure of 125 psi.

street elbow

An elbow with male thread on one end and female thread on the other end.

superheated steam

Steam at a higher temperature than that at which water would boil under the same pressure.

swing joint

A joint composed of screwed fittings and pipe to provide for expansion in pipe lines.

swivel joint

A joint employing a special fitting designed to be pressure-tight under continuous or intermittent movement of the machine or part to which it is connected.

tee

A fitting, either cast or wrought, that has one side outlet at right angles to the run.

travel

See **offset**.

union

A device used to connect pipes and usually consisting of three pieces: (1) the thread end fitted with exterior and interior threads; (2) the bottom end fitted with interior threads and a small exterior shoulder; and (3) the ring, which has an inside range at one end while the other end has an inside thread like that on the exterior of the thread end. Unions are extensively used because they permit connections with little disturbance of the pipe positions.

union ell

An ell with a male or female union at one end.

union joint

A pipe coupling, usually threaded, that permits disconnection without disturbing other sections.

union tee

A tee with a male or female union at one end.

welding-end valves

Valves without end flanges and with ends tapered and beveled for butt welding.

welding fittings

Wrought or forged-steel prefabricated elbows, tees, reducers, saddles, and the like beveled for welding to pipe.

welding neck flange

A flange with a relatively long neck beveled for butt welding to the pipe.

wiped joint

A lead pipe joint in which molten solder is poured upon the desired place, after adapting and fitting the parts together, and the joint is wiped up by hand with a moleskin or cloth pad while the metal is in a plastic state or condition.

wrought iron

Iron refined to a plastic state located in a puddling furnace. It is characterized by the presence of about 3% of slag irregularly mixed with pure iron and about 0.5% carbon.

wrought pipe

A term that refers to both wrought steel and wrought iron. *Wrought* in this sense means welded, as in the process of forming furnace-welded pipe from skelp or forming seamless pipe from plates or billets. The term *wrought pipe* is thus used to distinguish it from cast pipe. When wrought iron pipe is referred to, it should be designated by its complete name.

wye (Y)

A fitting, either cast or wrought, that has one side outlet at any angle other than 90 degrees.

—NOTES—

Appendix 1
THE PIPEFITTER

A pipefitter is a tradesperson who installs, assembles, fabricates, maintains, and repairs mechanical piping systems. Pipefitters usually begin as helpers or apprentices. Journeyman pipefitters deal with industrial/commercial/marine piping and heating/cooling systems. Typical industrial process pipe is under high pressure, which requires metal such as carbon steel, stainless steel, and many different alloy metals fused together through precision cutting, threading, grooving (Victaulic), bending, and welding. A plumber concentrates on lower-pressure piping systems for sewage and potable water (tap water) in industrial, commercial, institutional, and residential systems. Utility piping typically consists of copper, polyvinyl chloride (PVC), chlorinated polyvinyl chloride (CPVC), polyethylene, and galvanized pipe, which is typically glued, soldered, or threaded. Other types of piping systems include steam, ventilation, hydraulics, chemicals, fuel, and oil.

In Canada, pipefitting is classified as a compulsory trade, and carries a voluntary "red seal" interprovincial standards endorsement. Pipefitter apprenticeships are controlled and regulated by the province, and in some cases allow for advanced standing in similar trades upon completion.

In the United States, many states require pipefitters to be licensed. Requirements differ from state to state, but most include a 4- to 5-year apprenticeship. Union pipefitters are required to pass an apprenticeship test (often called a *turn-out exam*) before becoming a licensed journeyman. Others can be certified by NCCER (formerly the National Center for Construction Education and Research).

A pipefitter may install, assemble, fabricate, maintain, repair and troubleshoot pipe carrying fuel, chemicals, water, steam, or air in heating, cooling, lubricating, and various other processes piping systems. Pipefitters are employed in the maintenance departments of power stations, refineries, offshore installations, factories, and similar establishments by pipefitting contractors.

In North America, pipefitters are members of the United Association. Wages vary from area to area, based on demands for experienced personnel and existing contracts between local unions and contractors. The United Association is also affiliated with the piping trades unions in Ireland and Australia.

There is a difference between pipelayers and pipefitters is that pipefitters plan and test piping and tubing layouts, cut, bend or fabricate pipe or tubing segments, and join those segments by threading them, using lead joints, welding, brazing, cementing or soldering them together. They install manual, pneumatic, hydraulic and electric valves in pipes to control the flow through the pipes or tubes. These workers create the system of tubes in boilers and make holes in walls and bulkheads to accommodate the passage of the pipes they install.

Pipelayers, on the other hand, operate backhoes and trenching machinery that dig the trenches to accommodate the placement of sanitary sewer pipes and stormwater sewer drainpipes. They use surveyor's equipment to ensure the trenches have the proper slope and install the pieces of pipe in the trenches, joining the ends with cement, glue, or welding equipment. Using an always-open or always-closed valve called a tap, pipe layers connect them to a wider system and bury the pipes.

PIPEFITTER JOB EXAMPLES

Pipe Fitter—First Class Marine/Commercial Nuclear

Newport News Shipping Yard

Newport News, VA, USA

$29.50 hourly

Full-time

Journeyman Plumber/Pipefitter

LD Mechanical—Montrose, CO

Looking for responsible, motivated journeyman plumber for work around Montrose and Telluride, CO. Individual must have experience and knowledge of all plumbing/heating systems, be able to obtain a Colorado driver's license, and be clean, professional, and organized.

Job type: Full-time

Salary: $60,000.00/year

Required experience: Plumbing and heating: 4 years

Required license or certification: Plumbing license

Industrial Pipefitter

The Bell Company—Charlottesville, VA

The Bell Company is a large mechanical contractor that takes on some of the most complex projects in our East Coast Market. This is your chance to become a part of this dynamic team. As a pipefitter with The Bell Company, you will have the opportunity to use your skills to work on hospitals, laboratories, large industrial and institutional projects. Additional information on all our projects can be found on our Web site: www.thebellcompany.com.

PAY RANGE: $25–$27 hour for full journeyman pipefitter

THE BENEFITS: At the Bell Company we believe in rewarding members of our skilled crafts with more than just a competitive hourly rate.

We also offer:

Ongoing professional training and development

Opportunities for advancement

Potential career growth into our certified craft program with additional benefits of increased wages, paid holidays, and vacation accrual

Pipefitter Journeyman

Potent USA Corp.—Port Lavaca, TX

Minimum of 5 years' experience in the piping field. Read and understand field ISOs, layout, and field erection. Applicant to provide all tools for his craft.

Job type: Full-time

Salary: $34.00/hour

Required experience: Pipefitting: 5 years

Required education: High school or equivalent

Required license or certification: Driver's license

Required language: English

Required travel: 100%

Pipefitter Journeyman

Eastern Industrial Services of NY, Inc.—New York, NY

Install new heating, ventilation, and air conditioning systems

Must be able to read blueprints and build the waterside of HVAC systems from the provided plans and specs. Show up to jobsite on time every day ready to work.

Must be capable of building brazed, soldered, screw pipe, Victaulic, and welded piping systems

Adhere to all safety policies and procedures

Previous experience in steam fitting and/or pipefitting on an industrial or commercial scale

Familiarity with HVAC piping systems

Ability to handle physical workload

Strong problem-solving and critical thinking skills

Ability to plan and stay ahead of the job

Job type: Full-time

Salary: $15.00 to $40.00/hour

Experience: Pipefitting: 3 years (preferred)

License or certification: OSHA 10, G60, F60 (required)

Appendix 2
USEFUL DATA BANK

USEFUL FORMULAS

Water

Unless otherwise specified, data on specific weight and volume of water are usually at 60°F, although sometimes they are at 39.2°F, the point at which water has its greatest density. At 32°F, water weighs 62.42 pounds per cubic foot; at 60°F, 62.37 pounds per cubic foot; at 100°F, 62.00 pounds per cubic foot; and at 200°F, 60.13 pounds per cubic foot.

Steam

When water is raised to 212°F or its boiling point, 970 more Btu must be added per pound to convert the water to steam. Btu is the amount of heat required to raise the temperature of 1 pound of water 1°F at a temperature around 60°.

Temperature and Thermometers

The two principal thermometer scales are Fahrenheit (used in the United States) and Celsius (used in most countries and in most scientific work). In the Fahrenheit (F) scale, the freezing point is at 32° and the boiling point is at 212°, so that there are 180 divisions, or degrees, between them. In the Celsius (C) scale, freezing is at 0 and boiling is at 100°, with 100 divisions between them.

To convert degrees Celsius to degrees Fahrenheit, multiply the C reading by 9, divide by 5, and then add 32.

Example: What is 90°C in degrees Fahrenheit?

Answer: Multiply 90 by 9 and get 810. Divide by 5 and get 162. Add 32 and get 194°F.

To convert degrees Fahrenheit to degrees Celsius, subtract 32, multiply by 5, and divide by 9.

Example: What is 104°F in degrees Celsius?

Answer: Subtract 32 from 104 and obtain 72. Multiply by 5 and get 360. Divide by 9, and the answer is 40°C.

Horsepower

One horsepower is equivalent to 33,000 foot pounds per minute. A foot pound is the energy needed to raise 1 pound 1 foot vertically. The theoretical horsepower required to raise water a given height can be determined by multiplying the gallons to be raised per minute by 8.33. This multiplied by the vertical distance between the source and the point to which the water is to be raised gives the theoretical horsepower.

Tanks

To find the capacity of a cylindrical tank, square the diameter in feet, multiply by the length in feet, and then multiply by 0.7854. The result is the capacity in cubic feet.

Due to lack of barrel standardization, the number of gallons in a barrel varies for many different liquids. Consequently, it is safer to use cubic feet as a measure rather than barrels.

Example: What is the capacity of an 8-foot-diameter, 20-foot-long cylindrical tank?

Answer: 8 squared = 64
$$64 \times 20 = 1280$$
$$1280 \times 0.7854 = 1005.3 \text{ cu ft}$$

WEIGHTS AND MEASURES

Measures of Length

1 mile = 1760 yards = 5280 feet.

1 yard = 3 feet = 36 inches.

1 foot = 12 inches.

1 mil = 0.001 inch.

1 fathom = 2 yards = 6 feet.

1 rod = 5.5 yards = 16.5 feet.

1 hand = 4 inches.

1 span = 9 inches.

1 micro-inch = one millionth inch or 0.000001 inch.

1 micron = one millionth meter = 0.00003937 inch.

Surveyor's Measure

1 mile = 8 furlongs = 80 chains.

1 furlong = 10 chains = 220 yards.

1 chain = 4 rods = 22 yards = 66 feet = 100 links.

1 link = 7.92 inches.

Nautical Measure

1 league = 3 nautical miles.

1 nautical mile = 6,080.2 feet = 1.1516 statute miles. (The knot, which is the nautical unit of speed, is equivalent to a speed of 1 nautical mile per hour.)

One degree at the equator = 60 nautical miles = 69.096 statute miles (360 degrees = 21,600 nautical miles = 24,874.5 statute miles = circumference at equator).

Square Measure

1 square mile = 640 acres = 6400 square chains.

1 acre = 10 square chains = 4840 square yards = 43,560 square feet.

1 square chain = 16 square rods = 484 square yards = 4356 square feet.

1 square rod = 30.25 square yards = 27,225 square feet = 625 square links.

1 square yard = 9 square feet.

1 square foot = 144 square inches.

An acre is equal to a square the side of which is 208.7 feet.

Measure Used for Dia. and Areas of Electric Wires

1 circular inch = area of circle 1 inch in diameter = 0.7854 square inch.

1 circular inch = 1,000,000 circular mils.

1 square inch = 1.2732 circular inches = 1,273,239 circular mils.

A circular mil is the area of a circle 0.001 inch in diameter.

Cubic Measure

1 cubic yard = 27 cubic feet.

1 cubic foot = 1728 cubic inches.

The following cubic measures are also used for wood and masonry:
1 cord of wood = 4 × 4 × 8 feet = 128 cubic feet.

1 perch of masonry = 24.75 cubic feet.

Shipping Measure

For measuring entire internal capacity of a vessel: 1 register ton = 100 cubic feet.

For measurement of cargo:
Approximately 40 cubic feet of merchandise is considered a shipping ton, unless that bulk would weigh more than 2000 pounds, in which case the freight charge may be based upon weight.

40 cubic feet = 32.143 U.S. bushels = 31.16 Imperial bushels.

Dry Measure

1 bushel (U.S. or Winchester struck bushel) = 1.2445 cubic feet = 2150.42 cubic inches.

1 bushel = 4 pecks = 32 quarts = 64 pints.

1 peck = 8 quarts = 16 pints.

1 quart = 2 pints.

1 heaped bushel = 1.25 struck bushels.

1 cubic foot = 0.8036 struck bushel.

1 British Imperial bushel = 8 Imperial gallons = 1.2837 cubic feet = 2218.19 cubic inches.

Liquid Measure

1 U.S. gallon = 0.1337 cubic foot = 231 cubic inches = 4 quarts = 8 pints.

1 quart = 2 pints = 8 gills.

1 pint = 4 gills.

1 British imperial gallon = 1.2009 U. S. gallons = 277.42 cubic inches.

1 cubic foot = 7.48 U.S. gallons.

Apothecaries' Fluid Measure

1 U. S. fluid ounce = 8 drachms = 1.805 cubic inches = 1/128 U.S. gallon.

1 fluid drachm = 60 minims.

1 British fluid ounce = 1.732 cubic inches.

AREAS AND VOLUMES

Square	A = area. $A = s^2$ $A = \frac{1}{2}d^2$ $s = 0.7071 = \sqrt{A}$ $d = 1.414\sqrt{A}$
Rectangle	A = area. $A = ab$ $d = \sqrt{a^2 + b^2}$ $a = A \div b$ $b = A \div a$
Parallelogram	A = area. $A = ab$ $a = A \div b$ $b = A \div a$ Note that dimension a is measured at right angles to line b.
Right angled triangle	A = area. $A = \dfrac{bc}{2}$ $A = \sqrt{b^2 + c^2}$ $b = \sqrt{a^2 - c^2}$ $c = \sqrt{a^2 - b^2}$
Acute angled triangle	A = area. $A = \dfrac{bh}{2} = \dfrac{b}{2}\sqrt{a^2 - \left(\dfrac{a^2 + b^2 - c^2}{2b}\right)^2}$
Obtuse angled triangle	A = area. $A = \dfrac{bh}{2} = \dfrac{b}{2}\sqrt{a^2 - \left(\dfrac{c^2 + a^2 - b^2}{2b}\right)^2}$

Trapezoid	A = area. $A = \dfrac{(a+b)h}{2}$
Trapezium	A = area. $A = \dfrac{(H+h)a + bh + cH}{2}$
Regular hexagon	A = area; R = radius of circumscribed circle; r = radius of inscribed circle. $A = 2.598 = 2.598\ RE = 3.464$ $= 0.866\ R$
Regular octagon	A = area; R = radius of circumscribed circle; r = radius of inscribed circle. $A = 4.828 = 2.828 = 3.314\ a$ $R = 1.307\ r = 1.207\ 0.924\ R = 0.765\ R\ 0.828\ r$
Regular polygon	A = area n = number of sides $x = 360° \div n$ $B = 180° - x$ $A = \dfrac{nsr}{2} = \dfrac{ns}{2}\sqrt{R^2 - \dfrac{s^2}{4}}$ $R = \sqrt{r^2 + \dfrac{s^2}{4}}$ $r = \sqrt{R^2 - \dfrac{s^2}{4}}$ $s = 2\sqrt{R^2 - r^2}$
Circle	A = area; C = circumference; $A = 0.7854 d^2$ $r = C \div 6.2832 = 0.564\sqrt{A}$ $d = C \div 3.1416 = 1.128\sqrt{A}$ $C = 3.1416 d$

Frustum of cone	V = volume; A = area of conical surface. $V = 1.0472h(R^2 + Rr + r^2)$ $A = 3.1416\,s(R + r)$ $a = R - r \quad s = \sqrt{a^2 + h^2}$
Sphere	V = volume; A = area of surface. $V = 0.5236$ $A = 3.1416\,d^2$ $r = 0.6204 \times$ cube root of V
Spherical sector	V = volume; A = total area of conical and spherical surface. $V = 2.0944\,r^2h$ $A = 3.1416\,r(2h + \tfrac{1}{2}c)$ $c = 2\sqrt{h(2r - h)}$
Spherical segment	V = volume; A = area of spherical surface. $V = 3.1416B\,r - 9$ $A = 2\pi rh$ $c = 2\sqrt{h(2r - h)}$
Spherical zone	V = volume; A = area of spherical surface. $V = 0.5236\,h$ $A = 2\pi rh = 6.2832\,rh$
Barrel	V = approximate volume. If sides are bent to the arc of a circle: $V = 0.262\,h\,(2D^2 + d^2)$ If sides are bent to the arc of a parabola: $V = 0.209\,h\,(2D^2 + d^2)$

INTERNET LINKS RELATED TO PIPEFITTING

1. *Roth PEX-c Plumbing Systems Installation Handbook,* http://www.rothusa.com/PDF_Download_Files/Plumbing_Install_Manual.pdf.
2. Sharkbite Push-Fit Fittings, http://www.sharkbite.com/how-to/sharkbite-push-fit-fittings-how-to-install-pex-tubing-copper-tubing-and-cpvc-tubing/.
3. "How to Join Dissimilar Pipe Materials," *The Family Handyman,* https://www.familyhandyman.com/plumbing/plumbing-repair/how-to-join-dissimilar-pipes/view-all/.
4. *ASTM International Book of Standards,* https://www.astm.org/BOOKSTORE/BOS/index.html.
5. Plastic Pipe and Fittings Association, https://www.ppfahome.org/faq.aspx?prod=5.
6. Engineering Specifications for Viega PEX Tubing, https://api.ferguson.com/dar-step-service/Query?USE_TYPE=SPECIFICATION&PRODUCT_ID=2886838.
7. PEX Tubing Technical Specifications and General Installation Recommendations, https://www.pexuniverse.com/pex-tubing-technical-specs.
8. PEX Plumbing Information, http://www.pexinfo.com/.
9. ASPE—American Society of Plumbing Engineers, http://www.aspe.org.
10. IAPMO—International Association of Plumbing & Mechanical Officials, http://www.iapmo.org.
11. ICC—International Code Council, http://www.iccsafe.org.
12. NCSBCS—National Conference of States on Building Codes & Standards, http://www.ncsbcs.org.
13. ASSE—American Society of Sanitary Engineers, http://www.asse-plumbing.org.
14. ANSI—American National Standards Institute, http://www.ansi.org.
15. ASTM—American Society for Testing & Materials, http://www.astm.org.
16. CSA—Canadian Standards Association, http://www.csa.ca.
17. PPI—Plastics Pipe Institute, http://www.plasticpipe.org.
18. PHCC—The Plumbing-Heating-Cooling Contractors National Association, http://www.phccweb.org.
19. Sustainable Piping Systems, http://www.sustainablepipingsystems.com.
20. *Cast Iron Pipe and Fittings Handbook,* http://www.charlottepipe.com/Documents/CISPI_Tech_Man/CISPI_Tech_Man.pdf.

Index

Note: Page numbers followed by *f* refer to figures; page numbers followed by *t* refer to tables.

—NOTES—

—NOTES—

—NOTES—